# WAR
## TRANSFORMED

# WAR

## TRANSFORMED

The Future of Twenty-First-Century
Great Power Competition and Conflict

MICK RYAN

Naval Institute Press
Annapolis, Maryland

Naval Institute Press
291 Wood Road
Annapolis, MD 21402

Library of Congress Cataloging-in-Publication Data

Names: Ryan, Mick, 1969– author.

Title: War transformed : the future of twenty-first-century great power competition and conflict / Mick Ryan.

Description: Annapolis, Maryland : Naval Institute Press, [2022] | Includes bibliographical references and index.

Identifiers: LCCN 2021034928 (print) | LCCN 2021034929 (ebook) | ISBN 9781682477410 (hardcover) | ISBN 9781682477427 (ebook)

Subjects: LCSH: War—Forecasting. | Military art and science. | Strategy. | World politics—21st century.

Classification: LCC U21.2 .R92 2022 (print) | LCC U21.2 (ebook) | DDC 355.02—dc23

LC record available at https://lccn.loc.gov/2021034928

LC ebook record available at https://lccn.loc.gov/2021034929

∞ Print editions meet the requirements of ANSI/NISO z39.48-1992 (Permanence of Paper).

Printed in the United States of America.

30  29  28  27  26  25  24  23  22        9  8  7  6  5  4  3  2  1
First printing

*For the profession of arms*

# Contents

# Tables

# Acknowledgments

The story of this book did not begin when I put pen to paper (or fingers to keyboard) a couple of years ago. The true journey began in the late 1990s. After a succession of appointments as a combat engineer troop commander and engineer squadron second in command and a stint working operations in a joint headquarters, I found myself as the aide-de-camp of an Australian Army major general in our land headquarters in Sydney. I had not volunteered for the appointment and indeed had refused it several times. I had much to learn in those days.

It was a fortuitous appointment. I worked for a fantastic leader, Maj. Gen. Frank Hickling. A thoughtful commander and Vietnam veteran who understood how the world was changing, General Hickling reoriented our Army's focus from peacekeeping operations to land combat. He was eventually promoted and became our chief of army in mid-1998.

The time as aide-de-camp allowed me, for the first time, to see the bigger picture of an army as an institution. It reinforced my innate curiosity, and within a year I had published my first journal article. My selection to attend the U.S. Marine Corps Command and Staff College in 2001, followed by the School of Advanced Warfighting, supercharged my curiosity about our profession of arms and exposed me to new and exciting ideas about competition and conflict. It also allowed me to make a group of friends with whom I would serve in such places as Iraq, Afghanistan, and a cubicle in the basement of the Pentagon! We were the midranking officers of the post-9/11 generation, and many of us have remained in touch—I suspect we always will.

Since that time, I have benefitted from having many fine mentors and commanders who have influenced my thinking, my leadership style, and my approach to my various appointments in the Australian Army and beyond. Academics such as Toni Erskine, Mike Evans, Anne-Marie Grisogono, Frank Hoffman, David Johnson, Rob Johnson, Therese Keane, Tom Mahnken, Cathy Moloney, Williamson Murray, Linda Robinson, and Brendan Sergeant have all been generous with their time at points over the last two decades. In particular, the influence, counsel, inspiration, and friendship of Eliot Cohen have been enormous, and I owe him a debt of gratitude that these words will never fully convey.

I have been the beneficiary of the leadership, lessons, and mentorship of many fine Army, Navy, and Air Force leaders in both Australia and the United States, among them (leaving aside ranks) Dave Allen, Robert Bateman, J. B. Blackburn, Ken Bullman, John Caligari, Angus Campbell, Steve Day, Anthony "Fozzy" Forrestier, James Goldrick, John Hartley, Frank Hickling, David Hurley, Mike Krause, Pete Mangan, Gus McLachlan, Craig Orme, and Roger Noble.

The professional military education ecosystem that has grown up online over the past two decades has been a great supporter of my writing, and I am enormously appreciative of the relationships with the editors and staff of websites and blogs such as Modern War Institute, *War on the Rocks, Strategy Bridge, 3x5 Leadership, The Cove,* and others. Nate Finney, Steve Leonard (also known as Doctrine Man), and Joe Byerly in particular have been wonderful collaborators, friends, and advisors.

Over the past four years, I have cherished my time in command at the Australian Defence College in Canberra. The students provide me with endless inspiration—it is, after all, these students whom we seek to inspire and intellectually prepare for the coming fight. But the staff there are also absolutely first-class—they are one of the finest groups of military and civilian people with whom I have had the privilege to work. Every day, they are engaged in the crucial task of building the intellectual edge in our students to prepare them for the complex, changing, and challenging global security environment. In addition, I offer a very special shout-out to the team at our Australian Defence College library—I could not have done this without you.

Thank you to Glenn Griffith and the team at Naval Institute Press. Glenn was generous and courageous enough to take a chance on a foreign military officer who wanted to write his first book. The journey with the Naval Institute Press team has been an absolute pleasure, and I have learned much.

The Australian Army and its amazing people have given me more than I can ever repay. I have been honored to have been a small part of this unique and wonderful national institution. Being a part of the army and wanting to see it successful—and its people well led—were a significant impetus for writing this book.

Finally, through multiple operational tours, training exercises, work trips, house moves, overseas postings, absences on courses, and many other military demands, my daughters, Dana and Kara, and my wife, Jocelyn, have supported me. Jocelyn has carried the heaviest load through the majority of our marriage, has been the source of my best counsel, and is always my greatest cheerleader. This book and everything I have done over the last couple of decades would have been impossible without her—my best mate and the mother of our wonderful daughters.

This book may have taken two years to write, but it is a collection of what I have learned and observed over three and a half decades of soldiering in Australia, the Middle East, Southeast Asia, and the United States. As I approach the end of my military career, I trust that my legacy, in the form of this book, will be a useful contribution to the body of knowledge of our profession of arms.

# INTRODUCTION

## LADAKH, 15 JUNE 2020

Indian-controlled Ladakh and Chinese-controlled Aksai Chin lay opposite each other in the Himalayas. A line of control that neither India or China had ever agreed to roughly delineated their respective claims over the territory. For the Chinese, an agreement from November 1959 demarcated the line of control. For the Indians, a line of control agreed in September 1962 was the reference point.

Tensions in the Galwan Valley had been rising over the past several weeks. In May 2020 physical confrontations occurred between Indian and People's Liberation Army (PLA) troops near Pangong Lake. Both sides attributed the new unease to construction work on each side of the border to improve roads and airfields. Despite the tensions, several rounds of meetings had led to a deescalation of the situation between the two nuclear-armed powers and their respective border forces.

That all changed on the evening of 15 June 2020. The circumstances are still murky; China has consistently refused to offer a detailed account of the evening. What is known, however, is that late that evening, PLA soldiers crossed the line of control and attacked Indian soldiers bivouacked along desolate ridge lines that overlooked deep river gorges. Dozens, and perhaps hundreds, of soldiers from both sides fought a desperate battle on the dark, bitterly cold, and wind-swept ridges at an altitude of more than four and a half kilometers.

No gunshots were fired. Neither side used artillery or air support. Agreements struck between the two sides in 1996 and 2005 prohibited the firing of weapons. As one contemporary report noted, "Even though not strictly coded in any rules, officers said these practices have evolved over a period of time and have been firmed as part of a routine on the [line of actual control]. Since no round has been fired on the Sino-India border in Ladakh after 1962, these routines of not firing have been drilled into the soldiers. In such an environment, every other weapon short of firing has become acceptable to use for the soldiers."[1] The PLA attackers instead used barbed wire–wrapped sticks, iron rods, clubs, and rocks in their assault on the Indian positions. Engaged in desperate hand-to-hand combat, twenty Indian soldiers were either bashed to death or, more brutally, forced over ledges to fall hundreds of feet into freezing river gorges below. China acknowledged some casualties, and while it released no official figures, as many as forty PLA personnel are believed to have been killed.[2] Over the succeeding weeks, meetings between senior Chinese and Indian military leaders deescalated tensions. Reinforcements rushed to the area by both sides were eventually withdrawn. The status quo on the line of control was restored by late 2020.

Important lessons might be drawn from this instance of hand-to-hand combat on freezing high-altitude ridges in one of the most inaccessible regions in the world. First, not all warfare in the twenty-first century will feature advanced new weaponry, sensor networks, or classified systems. Humans will continue to compete and fight with the tools at hand. Warfare will employ a mix of old and new technologies—some ancient, and others almost magical in their technological sophistication.

Second, the Chinese Communist Party is challenging the status quo everywhere, even at the top of the world. Long-standing conventions in international behavior and use of military force are being challenged and breached across the Indo-Asian region. These tactics form part of China's approach in its strategic competition with the United States.

Third, seemingly minor incidents can occur out of nowhere, and large conflicts can develop from those small incidents if the right circumstances are present. We were fortunate that in Ladakh, these two nuclear-armed, highly populous nations stepped back from escalating the conflict. History shows that we have not always been so fortunate in the past—and may not be again in the future.

A final lesson is that surprise remains an enduring aspect of international relations, war, and competition. The nature of surprise means it does not just include sudden and unanticipated events; it also incorporates the types of conflict in which we might be engaged as well as technological breakthroughs and sudden natural events.

———

Humans keep getting surprised. As Roberta Wohlstetter writes in her classic study of surprise, *Pearl Harbor: Warning and Decision*, "The possibility of surprise at any time lies in the conditions of human perception and stems from uncertainties so basic that they are not likely to be eliminated, though they might be reduced."[3] Surprise is not going away, regardless of how technologically sophisticated humans become. Further, surprise creates shock in the mind of humans; this creates opportunities for our adversaries. However, we might act to understand or shape the range of events that might surprise us. In doing so, we can intellectually prepare our people for surprise and shock and ensure that they can adapt and mitigate the worst impacts when they are surprised.

Being surprised by an adversary is one of the most enduring elements of warfare. Indeed, achieving surprise is a principle of war in the military institutions of many nations.[4] Sun Tzu believed surprise was an indispensable tool for military commanders, writing that they must "attack where they [the enemy] are unprepared. Go forth where they will not expect it."[5] Two millennia later, the doctrine of military institutions in the United States, Russia, Britain, and Australia describes surprise as a key principle of war. From the Roman centurions' ambush by Germanic tribes in the Teutoburg Forest in AD 9 to General Maurice Gamelin's helplessness in the face of the German offensive in the Ardennes in France in 1940, surprise has resulted in defeat and catastrophic failure for some, glorious victory for others.

Regardless of how well connected we might be with sophisticated technologies or how much better we become at collecting and sharing information, surprise will continue to occur in the future—perhaps even more frequently. This is because we are witnessing transformations in military technology and operations at a rapid pace, driven by the changes in geopolitics, technology, and broader society.

Above all, the world is experiencing a revolution driven by the explosion of knowledge. Humans have always been curious beings. The history of humans demonstrates that we possess an insatiable need to know more about our world. But in the past two decades, the capacity to learn and share knowledge has expanded beyond any earlier capacity to do so. The connectivity of the Internet, the power to solve problems or find patterns through computational power, and the dawn of new algorithms that mimic some forms of human cognitive functions are driving this expansion. It is a new age of knowledge.

Some have called this new period an "era of accelerations" or a "great acceleration."[6] More often, it is described as the fourth Industrial Revolution. Geopolitics, demography, and the climate are all changing quickly, underpinned by rapid technological development across the fields of information technology, biotechnology, space, communications, and many others. The COVID-19 pandemic is providing a "supercharging" function in some of these areas, driving change at an unanticipated pace. This, combined with the impact of the ongoing disruptions in our environment, will fundamentally change the structure and resilience of societies and how they interact, as well as the way nations compete and make war.

The developments that have affected our societies are leading governments to reevaluate their conceptions of national security and national resilience. These reimaginings will disturb and unsettle how military institutions think about, deter, and prepare for future conflicts. They will also shake up how military organizations and their people conduct military operations. For some, these changes will be based on anticipating future challenges, assessing the changes in the surrounding world, analyzing the most effective strategies for military institutions, and then ruthlessly implementing these changes. A range of historical examples exists to guide policymakers and senior military leaders in this process.[7]

Unfortunately, in this new era, many military organizations will be forced to change because they failed to sufficiently appreciate the changes occurring around them, resulting in military failure. History provides endless examples of this. As Paul Kennedy notes in *The Rise and Fall of the Great Powers*, the great industrial and military developments of the nineteenth century ensured that "those powers which were defeated were those that failed to adapt to the *military revolution* of the mid-nineteenth century. . . . Grievous blunders were to be committed on the battlefield by the generals and armies of the winning side from time to time, but they were never enough to cancel out the advantage possessed in terms of trained manpower, supply, organization, and economic base."[8]

Nations and military institutions that have lagged behind their competitors have paid a price with the lives of their people and sometimes with their national sovereignty. As the Prussians in 1806, the Russians in 1905, the French in 1940, and the Iraqis in 2003 learned, the cost of paying insufficient attention to the competitiveness of one's military forces—intellectual, physical, and technological—is extremely high indeed.[9]

Military leaders must not be passive bystanders as these massive changes in technology, society, and geopolitics occur. Nations, military institutions, and strategists can do much to prepare and thrive in this new era. We know that there will be conflict in the future; the key aspects of its uncertainty include location and timing. However, we can and should be prepared. The historical record contains volumes of information and analysis on the industrial revolutions of the past quarter-millennium. There not only is a massive trove of histories on these periods, but there also are ample lessons that society and military institutions might use to inform the pathways ahead. Countries will have the opportunity to develop new strategies, build new organizations, refine operating concepts, and educate and develop their people to address the challenges of twenty-first-century warfare and strategic competition.[10]

The aim of this book is to explore what this revolution in the acquisition of knowledge and its resulting changes mean for military institutions. It examines the implications of ongoing developments in technology and society on war and competition. People, institutions, and ideas in military organizations are at the heart of this examination. The rapidly advancing technologies of the twenty-first century and their military applications can at times be overwhelming. However, technology is largely a level playing field. In the main, similar technologies are available to the major powers and even middle-sized nations (such as Australia). For twenty-first-century military institutions, building sources of advantage will include new technology, but it is not a complete solution to the many national security and military challenges of the coming decades.

Decades' worth of scholarship has explored past sources of military advantage. These historical studies have illustrated a key point: technologies alone do not provide military advantage. Technological development remains important and a worthy topic of examination, but it is how people combine technology with new ideas and new organizations that can provide a decisive advantage.

To truly revolutionize war, people and institutions must develop new methods of fighting and new ways of organizing; these are human, not technological, endeavors. As Audrey Kurth Cronin writes in *Power to the People*, "Despite the breathless claims of many technology promoters, what is crucial today for national security is not the transformative power of the new technologies, but the transformative power of human beings throughout the world to adapt them to unanticipated purposes."[11]

In his 2003 book *Battle: A History of Combat and Culture*, John Lynn argues against technological primacy in warfare. Instead, he proposes that those "who have tried to explain victory and defeat simply in terms of technological difference have generally erred. . . . The ability to maximize new technology depends very much on civil culture, such as religious, social, and political values."[12]

In a similar vein, strategist Thomas Mahnken has argued that "technology is developed and used by organizations, and the culture of those organizations has a great deal to do with which technologies are developed and how they are used."[13] One authority on military innovation, Williamson Murray, has proposed that the process of innovation and building effective military organizations is less about technology than about "a culture that allows innovation and debate unfettered by culture."[14] Another scholar of military studies and innovation, Stephen Biddle, notes that "technology alone is a poor predictor of capability. . . . It is the non-material component that is most influential."[15] And as the doyen of the strategic assessments community, the late Andrew Marshall, wrote in 1993, "The most important competition is not the technological competition. The most important goal is to be the best in the intellectual task of finding the most appropriate innovations in the concepts of operation and making organizational changes to fully exploit the technologies already available and those that will be available in the next decade or so."[16]

People, ideas, and organizations matter in military institutions and in strategic competition. They can be the difference between success and failure in war and in the great competitions between nations. As such, the core idea of this book is this: despite massive and ongoing advances in technology, it will be the combination of new ideas, new institutions, and well-trained and -educated people that will prove decisive for military organizations in twenty-first-century competition and war.

In this exploration, I can supply some but not all answers about preparing for future warfare and competition. There are areas where we might draw conclusions about potential pathways ahead; professional education comes to mind. But it is important that this book also pose questions. If we are to think in a truly strategic manner, our modus operandi must embrace an approach of continually diagnosing the environment in which we find ourselves and asking questions. No less than the great theorist and historian Carl von Clausewitz took this approach in writing *On War*. As Hew Strachan describes, "Clausewitz contradicts himself. But that is also *On War*'s strength, its very essence, and the reason for its longevity. It is a work in progress. Its author never stopped asking questions—not simply of his own conclusions but also of the methods by which he reached them."[17]

I have no pretensions that this book will reach anywhere near the grand scope or enduring greatness of *On War*. But I do hope that it serves to ask important questions and that it supplies some answers—that it fosters discussion and debate about war, competition, and the future of our profession of arms.

———

In his study of war and human conflict, *War in Human Civilization*, Azar Gat notes that "the solution to the enigma of war is that no enigma exists. Violent competition is the rule throughout nature. Humans are no exception to this general pattern."[18] For thousands of years of recorded history, humans have sought to impose their will on each other. Across this time, humans have expressed their creativity and brutality through the conduct of warfare against their fellow humans.

In the fifth century BC, Athenian general and historian Thucydides wrote of how "fear, interests, and honor" drove Athenian strategic decision-making.[19] Later, Clausewitz wrote in *On War* about the enduring nature and changing character of war. The enduring nature comprised the interaction of humans and the resulting friction, surprise, and uncertainty. Whether it is Spartans destroying the Athenian fleet at Aegospotami in 405 BC,[20] British and Australian soldiers campaigning on the Gallipoli Peninsula,[21] or U.S. soldiers undertaking stability operations and village security missions in Afghanistan, the essential and unchanged truth of war is the interplay of human beings.[22]

However, the character of war, or the means with which humans fight wars, has constantly evolved. The grand competition to possess the best weapons, achieve dominance over adversaries, and develop the best tactics and strategies for warfare has engaged the greatest minds throughout the ages. The ancient Greek historian Polybius wrote on Greek and Roman military tactics.[23] In the late fifteenth and early sixteenth centuries, Leonardo da Vinci wrote about the weapons of war and their employment on the land and at sea. While his sketches may have been beyond the technology of the time, his ideas reflected a desire to make war more efficient and to better protect soldiers and sailors from the dangers inherent in warfare.[24]

The character of war has also evolved through the application of new ideas to new technologies. The Greek hoplite phalanx was a core idea in ancient Greek military institutions.[25] Building on these formations, Philip of Macedon turned the shield wall of the Greek phalanx into a "spear wall" using six-meter-long pikes. Combining this with cavalry, light catapults, and more innovative "combined arms" tactics, he was able to use his new model army to conquer the southern Greek states by 338 BC.[26] Philip's son, Alexander, continued to develop the means of war, adding siege trains, staff, and a secretarial department. With this army, he conquered Asia Minor and advanced east as far as the Indus River.[27]

War continued to evolve in more modern times. The tactics of Frederick the Great, particularly his use of the oblique order and skillful rapid movement of troops, revolutionized warfare in the 1700s.[28] The French application of universal conscription through the levée en masse and Napoleon Bonaparte's formations of army corps had a profound effect on war in the early 1800s.[29] It drove reforms in tactics, armaments, and strategy in military institutions across the continent and beyond—by both those he had defeated in battle and those who watched from farther afield.

From the start of the U.S. Civil War in 1861 to the mid-twentieth century, a staggering amount of change occurred in warfare. From massed charges and nascent trench warfare, human conflict in the early twentieth century then also took to the skies and embraced whole-of-nation industrialized war. In the 1930s and 1940s, the scope of war broadened again to encompass the electromagnetic spectrum and eventually the atomic and space ages. In the past three decades, the depth and breadth of war have expanded further through the cyber realm.

Throughout it all, the combination of new ideas and organizations has exponentially improved the impact of new technologies. As multiple military historians have argued, where an optimum combination of new technologies, ideas, and organizations exists, the results often are what have come to be known as revolutions in military affairs.[30] Williamson Murray has described how military institutions embark on a revolution in military affairs to find improved ways to destroy their adversaries. It is a process that requires innovation across multiple military endeavors including tactics, people, organizations, technology, and doctrine. That said, however, institutions rarely set out to explicitly "revolutionize" warfare. Only in hindsight and in the wake of clear military victory does the revolution become clear.[31]

———

This book will examine how we can better anticipate the disruptions of the new era and how we might ensure we are prepared for the resulting changes in war, competition, and the capacity of military institutions to effectively achieve national security objectives. It reviews the adaptation needed for institutions, military ideas, and development of military personnel and concludes with the key unifying themes that will guide the application of military power in the twenty-first century.

Chapter one examines the drivers for change in war and competition. It first explores earlier periods of large-scale societal and technological transformation known as industrial revolutions and their impact on war. If we can understand how these industrial revolutions have changed war, national war-making capacity, and military organizations, we can then apply this knowledge to the emerging fourth Industrial Revolution. The chapter then reviews the status, and potential trajectories, of disruptive changes in the current era. Understanding the current and projected dimensions of this era and how it builds upon prior revolutions is an important element in thinking about the future of warfare.

Chapter two has warfare, the ultimate expression of human competition, as its focal point. War has been a preoccupation of humans for much of their existence; people and nations have waged war on each other for millennia. The different records of human warfare, in cave paintings, Greek urns, Roman statues and steles, and more modern writings, movies, and blogs, remind us that war is an enduring feature of human society. This chapter will identify the most important continuities and changes likely in future warfare, and then discuss the key trends in twenty-first-century war and competition. Without this understanding of war and how it is changing, exploring military power in the twenty-first century would not be possible.

Chapter three reviews the consequences for military institutions and their ideas in this new era. The chapter examines how the trends identified in the previous chapter will affect military institutions and their ideas at all levels of war. This examination of ideas and institutions is conducted through the lens of military effectiveness. Defining what makes a military organization effective is not simple. However, scholarship from the past three decades is instructive and is used to develop a picture of effective twenty-first-century military institutions and how they might operate.

The fourth chapter of the book concentrates on people. The profession of arms is unique in many ways. Samuel Huntington and Morris Janowitz examined the characteristics of the profession of arms in their ground-breaking studies in the 1950s.[32] These studies led to a broader acceptance of the military as a "profession" and helped shape institutional understanding of the imperatives of being the stewards of a profession. The fourth Industrial Revolution is already changing the nature of work across many industries. It is therefore highly relevant to explore how military "work"—and the recruiting, development, and leadership of military people—might also need to evolve. At heart, we must ensure military institutions are able to build for their people and their collective organizations the intellectual edge and physical capacity needed to thrive and succeed in a more lethal, ambiguous, and rapidly evolving environment.

The conclusion draws together the aspects of ideas, organizations, and people to build a plan of action for how nations might construct their future military power. It examines how war, as an enduring aspect of human existence, will continue to evolve in the twenty-first century and how nations and their military institutions must adapt. The chapter concludes with a series of propositions that are designed to assist leaders within military institutions and across a country's national security enterprise to develop the strategies and to prepare their people for the rigors of twenty-first-century conflict and competition. Rather than being new concepts, some of these propositions might instead reinforce old lessons that we may have forgotten. There are no easy solutions, but choices are available to those who have the courage and insight to take risk and invest in the right capabilities that will provide their nation with a competitive edge over the coming decades.

The primary audience for this book is current and future military leaders. Charged with the ethical application of violence to defend their nations and its interests, these leaders must sustain a commitment to learning, adaptation, and effective, values-based leadership over many decades. However, just like the rest of society, they have experienced and/or witnessed large transformations within their profession over the past thirty years.

Senior military officers of today will have served in a large-scale conventional force that faced off with the Soviet Union and then converted to train and deploy to defeat much smaller adversaries, with a focus on enforcing the peace within fractured nations. They will most likely have then deployed with an evolved and better-connected force that conducted operations in the Middle East and South Asia in the past two decades. These same officers must now adapt again to a world of revived strategic competition, where the main competitor is larger, richer, and more technologically advanced than any faced in living memory.

I have a personal stake in this. I have been a member of the profession of arms for more than three decades. I have had the honor of commanding military personnel from Australia and the United States and have witnessed first-hand the selfless and extraordinary dedication of military people and their families. That said, I have also observed many instances where military leaders, particularly at more senior levels from my generation, have eschewed the serious and necessary dedication to ongoing learning about war and about their profession. It is the reason that I have worked hard inside my own organization to change the culture of learning, to build adaptive capacity, to advocate for the development of an intellectual edge in our people, and to provide mentorship to the next generation of young military leaders.

Encouragingly, a new generation of junior officers and noncommissioned officers has established a global informal network through their work in social media and blogs. They have, in my view, reenergized the study of war, competition, and our profession and its future. It is this new generation of military leaders, more than any other audience, whom I wish to reach with this work.

Those who teach military personnel in training and education establishments, as well as members of the broader academic community, may also find much of interest in these pages. I also trust that this book informs our citizens of just how perilous the world has become. This is not meant to be a pessimistic assessment, but an evidence-based review of how technology and a new era of strategic competitions is driving different ways of thinking about military operations and institutions.

———

The journey of industrialization, scientific endeavor, and technical disruption over the last two centuries has laid a foundation for change in political systems, business, culture, the arts, and military institutions. Just as earlier industrial revolutions have changed war and military institutions, so too is it highly likely that the nascent fourth Industrial Revolution will do the same. For those unable to appreciate the changes around them or to muster the will or ability to sufficiently adapt to a changed environment, it is likely to pose great perils and potentially existential threats.

These changes in the function and means of war are of our own making. It is therefore within our gift to also mobilize the intellectual, physical, and moral forces to exploit the many opportunities of this revolutionary era. For those military institutions that are quick to anticipate, recognize opportunity, learn, and adapt, it will be an era of opportunity, prosperity, and security. Our current era has historical precedents from which we might learn.[33]

This book should aid military leaders to adapt to the current era to ensure their nations can evolve and thrive in the twenty-first century. They carry a heavy burden. As Williamson Murray writes, "War is neither a science nor a craft, but rather an incredibly complex endeavor which challenges people to the core of their souls. It is, to put it bluntly, not only the most physically demanding of all the professions, but also the most demanding intellectually and morally. The cost of slovenly thinking at every level of war can translate into the deaths of innumerable men and women, most of whom deserve better from their leaders."[34]

I have worked in this physically, morally, and intellectually demanding profession for nearly thirty-five years. The challenges of success in military operations and in producing effective strategies to support national leaders are a heavy burden carried by the leaders in this profession. The ongoing transformation of war makes this more difficult. While it is a Herculean task that those in the military and national security professions willingly embrace, the aim of this book is to make their burden more manageable.

# 1

## REVOLUTIONS and MILITARY CHANGE

### MIDTOWN MANHATTAN, 11 MAY 1997

The room, decorated like a study, had green walls and cream moldings and was adorned with several chest-high plants. But the most important element in it was a table holding a chess set. On one side, behind a small Russian flag, sat world champion chess master Garry Kasparov. Opposite him was IBM computer scientist Murray Campbell. He was not playing the game; his function was to move the pieces based on the instructions of a computer.[1]

The computer, one of the first generation of supercomputers, was housed outside the room, on the thirty-fifth floor of the New York Equitable Center skyscraper. Going into the contest, Kasparov was confident of once again beating the chess-playing machines. In 1989 Kasparov had defeated a Carnegie Mellon University computer called Deep Thought in a two-game series. Inspired by the efforts of the Carnegie Mellon team, IBM then developed the early version of Deep Blue.

In 1996 Kasparov had defeated a version of Deep Blue in a six-game tournament held in Philadelphia. While Kasparov had triumphed, the one game that Deep Blue had won in the match was enough to encourage the IBM team to continue developing the machine and its chess-playing capabilities. This work included creating new hardware to double the speed of the system and increasing its chess knowledge by adding "chess chips" that could recognize different positions and be more aware of chess concepts.[2] By 1997 the Deep Blue machine could explore up to 200 million chess positions per second.[3] It was also programmed with small tricks, particularly in how long it took to make moves, to try to shock or unsettle Kasparov psychologically.[4]

In this sixth and final game of the May 1997 series, Kasparov was under pressure. He had won one game, the computer had won one game, and the rest had been a draw. But the strain was starting to show in both Kasparov's game play and his demeanor. On its eighth turn, Deep Blue played an aggressive move in which it sacrificed a knight. From this point, Kasparov never recovered, and he lost in nineteen moves.[5]

IBM had finally achieved its aspiration of winning a series of games against a reigning world chess champion. A human being had his livelihood, his life's work, disrupted by a machine that gave the appearance of outthinking and outplaying him in chess. The rigid commitment of the machine and its cold logic had triumphed over a human. History will not record Garry Kasparov as one of the first human beings to be replaced by a machine. But he may be one of the first whose cognitive functions were outmatched by a machine that gave the appearance of intelligence. While we understand that machines do not "think" and do not possess "intelligence," the IBM computer that bested Garry Kasparov in 1997 demonstrated some of the cognitive functions that are traditionally associated with human beings.

The IBM Deep Blue victory—a chess grand master outplayed by a machine—sparked controversy, acrimony, and wonder. It began a disruption of white-collar workers by providing the wherewithal to undertake more complex mathematical calculations more quickly than humans could and to automate many office functions. But in some respects, this was just another step in the history of machines that had disrupted the manufacturing and agriculture industries over the previous two centuries.

## REVOLUTIONS PAST

Over the past 250 years, several pulses of innovation, classified as industrial revolutions, have occurred. These revolutions produced technologies and other changes that have transformed military affairs.[6]

The original Industrial Revolution, which started in England in the 1760s, was powered by steam and informed by the telegraph machine, and it significantly increased industrial productivity.[7] Coal mining and the mechanized production of flour and textiles resulted in a massive increase in the output of workers employed in mills. Before 1820 economic growth over the preceding centuries had been slow. After 1820 England, many nations in Europe, and the United States became vastly more prosperous.[8]

This first Industrial Revolution had significant impacts for human societies. A transition to urban living was a major element. Industrialization attracted workers by the thousands to the urban centers where factories were located. For the first time, large numbers of people now worked outside the local area of their homes. The pace of work for people employed in large factories also changed. Instead of relying on the speed of horses, people worked at the speed of machines.

The conduct of war was also transformed by this industrial revolution. Military institutions could now exploit a range of tactical and operational capabilities that used new technologies. The railroad, the steamboat, the telegraph, and mass-produced longer range weapons combined to change the face of warfare. Due to more efficient national mobilization, military commanders also were able to lead larger armies. They were able to lead these armies on campaigns that stretched farther across the landscape and for longer periods. The mass manufacture of weapons, munitions, and supplies underpinned better support to armies and navies. At the same time, military commanders (and their national leaders) now received unprecedented levels of information more rapidly than any previous generation.

Historians Williamson Murray and Wayne Hsieh have characterized this period as a "military-social revolution" during which large-scale change that occurred in states and societies also drove transformation in their military institutions. The first of these military-social revolutions was the creation of the modern nation-state and bureaucratic, disciplined military organizations. Successive military-social revolutions in the nineteenth century, which saw the development of modern war in the wake of the U.S. Civil War and the Franco-Prussian War, shaped the conduct of military operations into the twentieth century.[9]

The first Industrial Revolution provided the foundations for subsequent technical revolutions that also would have a profound influence on societies, as well as on war and competition.

### The Second Industrial Revolution

The next surge of industrial transformation, which occurred in the last third of the nineteenth century and early part of the twentieth century, is sometimes characterized as the greatest technical discontinuity in history.[10] This new era was launched by fundamental breakthroughs in science, including the first practical designs for open-hearth steelmaking furnaces (1867) and dynamos (1867), the formulation of the second law of thermodynamics (1867), and the development of dynamite (1868). Heinrich Hertz's discovery of electromagnetic waves in 1886 led to wireless radio communications and, later, television. Advances in combustion engines, the first continuously moving assembly line, the coiled tungsten filament (still found in light bulbs), manned flying machines, and large-scale electricity generation changed industry and broader society.

Accompanying this massive expansion in scientific knowledge and improved manufacturing was a rapid absorption of these technologies in society. Many innovations from the three decades before World War I were patented and commercialized within months (such as the light bulb and telephones) or just a few years (such as petrol-fueled automobiles).[11]

The technologies that emerged from this second Industrial Revolution also underpinned the next great transformation in military affairs. The power of flight, internal combustion engines, wireless communications, radar, and electrically powered factories were all part of this important story. Mass production and the expanded capacity of nations to exploit their economies for warfare combined to produce disruptive change in how military organizations prepared for and conducted military operations.

But technologies were just the foundation; it was their combination with new ideas, new organizations, and new approaches to production and economic management that produced truly revolutionary developments. By combining new technologies with new organizational arrangements and new warfighting ideas, military organizations could fight differently and move more rapidly. They could also mobilize more quickly, adopt standardized equipment and organizations, and be supported by strategic industries with a larger quantity and wider range of military supplies.[12]

The technologies of the second Industrial Revolution found their ultimate military expression in World War II. While there were changes to military institutions and the conduct of warfare in the 1920s and 1930s, the second Industrial Revolution reached its full manifestation in the 1940s.[13] New ideas of warfare and combat were combined with new technologies such as the machine gun, tanks, radios, submarines, aircraft carriers, and aircraft to produce combined arms warfare on the land, carrier and amphibious warfare at sea, and the firebombing of cities from the air. The use of atomic bombs against the Japanese cities of Hiroshima and Nagasaki in 1945 was made possible by the myriad of scientific and industrial advancements of the second Industrial Revolution. These new weapons heralded the potential for even more devastating wars.

Czech-Canadian scientist Vaclav Smil has described the second Industrial Revolution as an "astonishing concatenation of epochal innovations that were introduced and improved during the two pre-[World War I] generations and whose universal adoption created the civilisation of the 20th century."[14] The technological wonders of this period improved industrial productivity while they also enhanced the prosperity and well-being of people across Europe, Britain, North America, and beyond. Industrial and agricultural output surged, transportation networks expanded, and the capacity to share information more quickly through the telegraph and then the telephone saw the development of an international trading system that decades later would be termed globalization.

### A Third Revolution

The last four decades of the twentieth century have been described as another revolutionary period. This new revolution was different in character from its predecessors. It was less about horsepower and more about the power of vacuum tubes and silicon. Whereas the previous two industrial revolutions had improved humanity's physical capacity, the third industrial revolution augmented humankind's capacity to communicate, generate, and share knowledge.

Like earlier revolutions, however, this was a period of surging techno-logical developments founded on the discoveries of previous decades. Developments in computing and transistors in the 1940s and 1950s, the invention of microprocessors in the late 1960s and early 1970s, the U.S. Defense Department's construction of the Advanced Research Projects Agency Network—ARPANET—from 1969 to 1977, and expanded access to the Internet in the 1990s underpinned an explosion in information access, consumer communications, and wide availability of cheap comput-ing power and digital devices. The third Industrial Revolution was an era of increasingly more affordable computational power for businesses, govern-ments, and individuals. It also featured the birth and massive growth of the Internet[15] and the expanded use of space-based capabilities such as precision navigation and timing, communications, and space-based sensing.

The era also resulted in crucial advances in military technologies, includ-ing vastly improved battlefield connectivity and awareness, precision weap-ons, dense intelligence, surveillance, and reconnaissance networks, and more streamlined and connected military logistics. New technologies and a strate-gic competition with the Soviet Union drove a series of innovations in how Western military forces thought about and organized for war. Joint opera-tions and nuclear deterrence theory are but two of the new ideas that emerged and were greatly refined in this period. The third Industrial Revolution also resulted in space travel, satellites, precision weapons, as well as joint and cyber warfare.

## A NEW REVOLUTION

The world is now on the threshold of a new pulse of rapid change founded on global connectivity, silicon-based artificial intelligence, biotechnology, and advances in the ability and affordability of different robotic systems. This so-called fourth Industrial Revolution is disrupting business, entertain-ment, communications, transportation, and national economies.[16] Like ear-lier revolutions, it is affecting societies around the globe.

What differentiates this new revolution is not physical production, but rather the production of knowledge. It is a transformation in the cognitive capacity of human beings—their capacity to generate and share knowledge and then apply that knowledge to new and old human endeavors in society, commerce, science, and national defense. This revolution is displacing not just blue-collar workers; it is also making redundant many forms of white-collar work such as accounting and administration.

The unprecedented speed at which some of the change—particularly in technology—is occurring magnifies its impact while also making surprise more likely in the future. Consequently, nations will require long-term agile strategies complemented by an intrinsic capacity to rapidly adapt at all levels to both anticipated and surprising changes in this new environment.[17]

Adding fuel to the fire of this new revolution is the COVID-19 pandemic, which has forced governments, corporations, and societies to reevaluate many aspects of their national approaches to resilience and sovereignty. The catastrophic events of 2020 and 2021 revealed the shortfalls of an over-reliance on market forces for the supply and stockpiling of essential manufacturing capacities, particularly pharmaceuticals and personal protective equipment. While not every nation can manufacture every commodity that it consumes, the concentration of manufacturing power within China for key products will undoubtedly be closely scrutinized in a post–COVID-19 world. Momentum is gathering within large and mid-size economic powers such as the United States, the United Kingdom, Japan, and Australia to increase autonomy and enhance their national resilience.

## INNOVATION: PULSES AND PAUSES

Vladimir Lenin reportedly observed that "there are decades where nothing happens; and there are weeks where decades happen." Industrial revolutions and the surges of change they represent appear to fit this aphorism. But the events occurring during industrial revolutions do not tell the entire story of human innovation. A quick look at the accepted time periods of the four industrial revolutions also reveals significant periods of time between revolutions. During these pauses, innovation does not cease entirely; it just occurs at a slower pace. The following section briefly examines how innovation can progress rapidly for short periods of time and why it can be a bewildering experience for those caught in the midst of such an era.

In his 1979 book, *Stalemate in Technology*, economist Gerhard Mensch undertook a study of two hundred years of innovation. He found that innovation does not occur continuously. Instead, in exploring hundreds of innovations over the period 1740 to 1960, Mensch found a "dramatic alternation between periods of innovative abundance and innovative scarcity," which he termed the "discontinuity hypothesis."[18]

Mensch's "dramatic alternation" aligns with the pulses of innovation that occur during industrial revolutions—pulses that are sometimes accompanied by a sense that change is not just occurring more rapidly, but also accelerating. In a 2003 interview, Chris Meyer and Ray Kurzweil discussed a range of issues related to the pace of development in several human endeavors, including technology. Kurzweil noted that

> While acceleration was there 500 years ago, it was at that point of an exponential where it looked like a flat, horizontal line. The first time the rate of change was really disrupted, was with the weavers in the English textile industry, who had had a weaving guild that had been passed down for centuries through their families. We're entering an era of acceleration. The models underlying society at every level, which are largely based on a linear model of change, are going to have to be redefined. . . . The twenty-first century will be equivalent to 20,000 years of progress at today's rate of progress.[19]

In 2016 Thomas Friedman examined this accelerating pace of change in his book *Thank You for Being Late*. Friedman explores the impact of three interacting themes—technology, the market, and climate change—noting that "the market, mother nature, and Moore's Law together constitute the Age of Accelerations in which we find ourselves."[20]

Another theme of Friedman's book is that the current era has become quite dizzying for many people. Most of us hear about breakthroughs in biotechnology, robotics, artificial intelligence, and other areas of scientific endeavors, but we are unsure about where these advances will take us. In previous revolutions, the pace of absorption of new technologies was such that there was time to develop new laws, societal norms, and other systems that allowed people to understand and use the new technologies. That is not the case now. As Friedman notes, "Legislatures are scrambling to keep up, tech companies are chafing under outdated and sometimes nonsensical rules, and the public is not sure what to think."[21]

This sense of accelerating change also applies to earth sciences and climate studies. The authors of a 2004 book, *Global Change and the Earth System*, examined the evidence for the impact of human activities on a range of earth systems. While the authors—experts in a variety of earth sciences—expected to see evidence of a "growing impact of the human enterprise on the earth system," they discovered that there had been a change in the magnitude of this impact since the middle of the twentieth century.[22] The authors noted that "one feature stands out as remarkable. The second half of the twentieth century is unique in the entire history of human existence on Earth. Many human activities reached take-off points sometime in the twentieth century and have accelerated sharply towards the end of the century. The last 50 years have without doubt seen the most rapid transformation of the human relationship with the natural world in the history of humankind."[23] This trend in accelerating change since the 1950s was also highlighted in the most recent Intergovernmental Panel on Climate Change, which notes that "warming of the climate system is unequivocal, and since the 1950s, many of the observed changes are unprecedented over decades to millennia. The atmosphere and ocean have warmed, the amounts of snow and ice have diminished, and sea level has risen."[24]

Finally, the theme of accelerating change has been explored from a military perspective. In 2005 Max Boot wrote about various transformations in the conduct of military operations over the past several centuries. In *War Made New*, Boot's focal point was the impact of technology on war since 1500. While his book contains many valuable insights, one of the more useful is that "innovation has been speeding up. It took more than 200 years for the gunpowder revolutions to come to fruition, 150 years for the first Industrial Revolution, 40 years for the second Industrial Revolution and 30 years for the Information Revolution. Keeping up with the pace of change is getting harder, and the risks of getting left behind are rising."[25]

Military and national security institutions, absorbed by the wars in Iraq and Afghanistan since the turn of the century, have only in the past decade been examining this phenomenon of accelerating change as a distinct aspect of the future environment. It was not identified as a trend in the 1997 U.S. National Intelligence Council publication, *Global Trends 2010*. The accelerating pace of technology received only a single mention in the council's 2000 report but was then featured in the December 2004 report, *Mapping the Global Future*. In the 2012 publication *Sustaining U.S. Global Leadership: Priorities for 21st-Century Warfare*, the U.S. Department of Defense proposed that accelerating change was a key element of the global environment in which the U.S. military would be operating in the future.[26]

Accelerating change is a theme in the 2017 National Intelligence Council report on global trends. As the report notes, "Artificial intelligence and robotics have the potential to increase the pace of technological change beyond any past experience, and . . . may be outpacing the ability of economies, societies, and individuals to adapt."[27] A more recent publication from the U.S. Army's Training and Doctrine Command has described the period out to 2035 as an "era of accelerated human development."[28]

Therefore, many government and business entities have accepted that we are experiencing a period of rapid and accelerating change in technology, work, and society. One only needs to read the massive number of reports from business consultancies or public statements by different government agencies to get the message. However, does the evidence support the hypothesis of Kurzweil, Friedman, Boot, and others that we are in a period of accelerating and historically unprecedented change?

Exploring the world at the beginning of the twentieth century in his book *The Vertigo Years*, author Philipp Blom notes that "speed and exhilaration, anxiety and vertigo were recurrent themes of the years between 1900 and 1914, during which cities exploded in size and societies were transformed, mass production seized hold of everyday life, newspapers turned into media empires. . . . Rapid changes in technology, globalization, communication technologies and changes in the social fabric dominated conversations and newspaper articles; then as now, the feeling of living in an accelerating world, of speeding into the unknown, was overwhelming."[29] As in our own era, this time of rapid change engendered among some people fears of the brisk developments in technology and society.

Edward Bryn also felt a sense of rapid and accelerating change in 1900. He noted in *The Progress of Innovation in the Nineteenth Century* that "whatever the future centuries may bring in new and useful inventions, certain it is that the Nineteenth Century stands pre-eminent in this field of human achievement, so far excelling all other like periods as to establish on the pages of history an epoch as remarkable as it is unique."[30]

In *Stalemate in Technology*, Gerhard Mensch critiques the notion of accelerating change in technological developments. In addition to his discontinuity hypothesis, he finds that "basic innovations do not generally appear at a progressively faster rate. Only sometimes do basic innovations emerge at a rapidly accelerating pace, namely after the initial retardation has increased the pressure for realization to such a degree that it produces a surge of innovations."[31] He posits that this is a rare and exceptional situation, however. Possibly, we now find ourselves in one of these rare periods of surging innovation.

Another assessment of the idea of accelerating change and of the fourth Industrial Revolution concept more broadly comes from Tim Unwin, who critiques the overwhelming technological bias of those who are key "thought leaders" in the current revolution. He also posits that "technology itself does not change anything. Technology is designed by people for particular purposes that serve specific interests. It is these that change the world, and not the technology." Unwin explores the fundamental basis for the term "industrial revolutions" and questions whether they are actually just extensions of innovation and change from previous eras. Finally, Unwin describes the fourth Industrial Revolution as an "elite view of history," a self-fulfilling prophesy and a period where the heroes are largely male.[32]

In 2016 Elizabeth Garbee critiqued the notion that the current era represents something new in human history. Instead, Garbee proposes that the current era is only repeating themes of previous industrial revolutions. These recurring themes include the convergence of new and evolved technologies that "change the scale, scope, and complexity of our collective intellectual landscape." Garbee, however, accepts that rapid pace of change but notes that "even exponential growth rates, often cited as the defining feature of this so-called fourth Industrial Revolution, are nothing special—they occur any time a system grows at a constantly fractional rate."[33]

A final, and more brutal, critique of accelerating change in the current era is provided by Vaclav Smil. Exploring the second Industrial Revolution in his book *Creating the Twentieth Century*, Smil does not dispute that the twenty-first century is seeing rapid changes in technology and society. He also accepts that some technological trends such as Moore's Law have seen exponential growth. But on the matter of accelerating change, he is clear: "Those commonly held perceptions of accelerating innovation are ahistorical, myopic perspectives proffered by zealots of electronic faith."[34]

———

What insights might be extracted from this examination of innovation pulses, pauses, and acceleration?

First, human existence is punctuated by surges in innovation every few decades. As George Orwell wrote in 1942, "Every now and again, something happens—no doubt it's ultimately traceable to changes in industrial technique . . . and the whole spirit and tempo of life changes, and people acquire a new outlook which reflects itself in their political behavior, their manners, and everything else."[35] Those who live through these surges often feel that change is not only rapid and bewildering but also accelerating in pace. This has occurred multiple times in the past two hundred years. This is exemplified when Blom describes how people living in the decade before World War I believed that they were "living in an accelerating world, of speeding into the unknown."[36]

Second, not all important innovations occur during the surges, or what we call industrial revolutions. As Chris McKenna has noted, "What happens when there is a pause? Pauses are as important as periods of acceleration."[37] There is historical evidence to support this. For example, the machine gun—so important in World War I—had its origins in the period between the first and second Industrial Revolutions. Computers appeared between the second and third Industrial Revolutions. Importantly, major events such as wars have also taken place during the pauses between industrial revolutions. The U.S. Civil War, the Franco-Prussian War, and World War II are all examples.

Third, as the pace of change increases during innovation surges, humans and the organizational approaches that underpin society must adapt more rapidly to technological developments. The ideas that are developed, tested, and used in business, government, academia, and other human endeavors must also evolve more rapidly during these pulses of innovation.

Fourth, while surges in innovation enable new ways of doing things we are already doing, new and novel challenges and industries can also emerge. This drives humans to think anew about problems and to develop new and more efficient organizations to implement these ideas. New ideas and institutions also drive change in how humans are trained and educated. This can then nurture the development of more strategic approaches to problem-solving to complement preexisting local solutions.

Fifth, since the 1760s, industrial revolutions have provided a foundation for a series of military transformations that disrupted military institutions and their ideas while reshaping competition and the conduct of warfare.[38] These transformations have often been described as revolutions in military affairs. War has continued to evolve, through to the present, shaped by technological developments, innovative ways of employing new technologies, and the industrialization of producing military equipment and supplies.

Sixth, while industrial revolutions have changed the material or weapons of war, they have heralded a more profound impact on the cognitive aspects of human conflict. A new tool in the hands of an unthinking automaton is largely useless. A new tool matched to a new idea can be unbeatable. Therefore, we must use caution when exploring the future dimensions of human conflict so we are not seduced by the allure of new wonder technologies being promised as game changers. New weapons rely on new ideas and organizations to reach their potential.

With few exceptions, most historical game changers in the military profession have been cognitive, not physical. Whether we consider the tactics and training of Frederick the Great, the operational and strategic logistics of the

Union Army in 1861–65, or the breakthroughs in combined arms, amphibious operations, or carrier operations in the 1920s and 1930s, it is the wit of humans and not their muscle that has been—and will continue to be—decisive. The material of war will remain an important success factor, but the brains and the will of humans will continue to decide the outcome of war.

From trench warfare to aerial conflict, combined arms operations on the land and carrier warfare at sea, the development of nuclear deterrence theory of the 1950s and 1960s, and the application of AirLand Battle against the Iraqi army in 1991, the combination of new ideas and organizations has exponentially improved the impact of new technologies. As multiple military historians have argued, where there is an optimum combination of new technologies, ideas, and organizations, transformation in military activities has resulted.

## CONTEMPORARY DRIVERS OF DISRUPTION

We are on the cusp of a new pulse of innovation. Like earlier innovation surges, this period is likely to play out over several decades before developments in technology and society slow down again. The current pulse of innovation will demand that military institutions reconceive their threat assessments, plan how to evolve, and assess how much risk they can take in rebalancing resources to invest in people, equipment, doctrine, and organizations.

However, before reviewing the kinds of changes necessary for military institutions, it is important to understand the character of the disruptive changes that are occurring around us. As Michael O'Hanlon has noted, "Technological change of relevance to military innovation may be faster and more consequential in the next 20 years than it has proven to be over the last 20. Notably, it is entirely possible that the ongoing, rapid pace of innovation may make the next two decades more revolutionary than the last two."[39]

The next section explores the key elements of disruption in the global security environment that will shape the future of military activities. Understanding the trends that are driving disruption provides a critical foundation for preparing military people to think, plan, compete, cooperate, and fight in the twenty-first century.

Drivers of change in the global environment and the security environment are studied widely in academia, the business community, and military institutions. In academia, institutions such as the Atlantic Council and RAND have programs that study key strategic drivers and their impact on international relations and the security environment.[40] Commercial entities such as McKinsey and KPMG explore potential future scenarios and develop insights for business leaders. These insights are often focused on the impact of new and evolving technologies on workforces and business strategy.

In military organizations around the world, bespoke institutions are dedicated to the study of strategic trends and future scenarios to inform national defense planning. These insights, while sometimes classified, are often published in standalone reports, as journal articles, or as part of published national defense assessments and strategies. Examples of standalone reports include the *Future Trends* series by the United Kingdom's Development, Concepts and Doctrine Centre and the U.S. National Intelligence Council *Global Trends* reports (see table 1.1).

Military organizations that wish to remain relevant in the twenty-first century must continue to develop their understanding of this future security environment. If they have an idea of their potential operating environments, it will be possible to design and construct military institutions that are prepared for this environment. There are no guarantees of success. But building knowledge about different potential futures allows large institutions to tailor their approaches for future effectiveness.

A survey of the broad array of publications and reports over the last decade shows some notable trends, which might also be described as principal disruptors. They include geopolitics, demographics, technology, and climate and natural threats.

## DISRUPTOR ONE: NEW-ERA STRATEGIC COMPETITION

In "The End of History?" an article published in *The National Interest* in 1989, Francis Fukuyama proposed that the twentieth century appeared at its close to be "returning to where it started . . . to an unabashed victory of economic and political liberalism." His central argument was that the spread of liberal democracy after the fall of the Soviet Union might indicate the end of humanity's socio-cultural evolution with "the universalization of Western liberal democracy as the final form of human government."[41]

## Table 1.1. Key Findings of Strategic Environment Scans

| Organization | Publication | Drivers for Change |
|---|---|---|
| **Think Tanks and Academia** | | |
| RAND Corporation | *Discontinuities and Distractions: Rethinking Security for the Year 2040* [1] | Demographics and dependency; weather and climate; information insecurity; changes in the liberal international order; changes in warfare |
| Atlantic Council | *Global Risks 2035 Update: Decline or New Renaissance?* [2] | Geopolitics (likelihood of state-on-state war, fracturing of global system, deglobalization); demography (poverty); climate change; technology |
| Center for Strategic and International Studies | *Defense 2045: Assessing the Future Security Environment and Implications for Defense Policy Makers* [3] | Demographics and urbanization; economics and national power; power diffusion and the global order; technology; connectedness; geopolitics |
| Center for Strategic and Budgetary Assessments | *Why Is the World So Unsettled: The End of the Post–Cold War Era and the Crisis of the Global Order* [4] | Geopolitics (erosion of U.S. and Western primacy, new great power competition, return to ideological struggle); decline of Western military power; intensification of global disorder |
| Commonwealth Scientific and Industrial Research Organisation | *Australia's National Outlook 2019* [5] | Rise of Asia; technological change; climate change and environment; demographics; trust; social cohesion |
| **Business Community** | | |
| KPMG | *Future State 2030: The Global Megatrends Shaping Governments* [6] | Demographics; rise of the individual; enabling technology; economic interconnectedness; public debt; economic power shift; climate change; resources stress; urbanization |
| McKinsey | *Economic Conditions Snapshot, March 2020: McKinsey Global Survey* [7] | Coronavirus; geopolitical instability; trade conflicts; financial market volatility; transitions of political leadership |
| **Government and Military Studies Centers** | | |
| UK Development, Concepts and Doctrine Centre | *Global Strategic Trends: The Future Starts Today* [8] | Environment and resources; human development; economy, industry, and information; governance; conflict and security |
| National Intelligence Council | *Global Trends: Paradox of Progress* [9] | Aging; shift in global economics (nature of work); accelerating technological change; ideas and identities; governance challenges; the nature of conflict; climate change and the environment |
| Government of France | *Defence and National Security Review 2017* [10] | A challenged international system (multilateral order questioned); multiple weaknesses aggravating crises (demography, climate change, energy, sanitation, crime); disruptive technology; harder and more disseminated threats |

| Organization | Publication | Drivers for Change |
| --- | --- | --- |
| Government of Russia | *National Security Strategy* [11] | Changing (polycentric) model of world order and instability; international competition encompassing social and technological elements; role of force in international relations important; U.S. global strike capability; nuclear proliferation; demographic challenges |
| Chinese Communist Party | *China's National Defense in the New Era* [12] | Strategic competition; information society; cultural diversification; multipolar world; global military competition |
| U.S. Office of the Director of National Intelligence | *Worldwide Threat Assessment of the U.S. Intelligence Community* [13] | Cyber; online influence; weapons of mass destruction; disruptive technologies; space systems; transnational organized crime; economics and energy; human security (health, displacement, freedom, climate change) |
| Strategic Studies Institute | *Strategic Landscape 2050: Preparing the U.S. Military for New Era Dynamics* [14] | Demographic dividends and liabilities; environmental risks and breakthroughs; uneven socioeconomic and political transitions; technological disruption; military revolutions; regional, technological and military races |

1. Andrew Hoehn et al., *Discontinuities and Distractions: Rethinking Security for the Year 2040* (Santa Monica, CA: RAND Corporation, 2018).
2. Matthew Burrows, *Global Risks 2035 Update: Decline or New Renaissance?* (Washington, DC: Atlantic Council, 2019).
3. David Miller, *Defense 2045: Assessing the Future Security Environment and Implications for Defense Policy Makers* (Washington, DC: Center for Strategic and International Studies, September 2015).
4. Hal Brands and Eric Edelman, *Why Is the World So Unsettled: The End of the Post–Cold War Era and the Crisis of the Global Order* (Washington, DC: Center for Strategic and Budgetary Assessments, 2017).
5. Commonwealth Scientific and Industrial Research Organization, *Australia's National Outlook 2019* (Canberra: Commonwealth Scientific and Industrial Research Organization, 2019).
6. KPMG, *Future State 2030: The Global Megatrends Shaping Governments* (Toronto: KPMG International, 2013).
7. Alan FitzGerald et al., *Economic Conditions Snapshot, March 2020: McKinsey Global Survey* (New York: McKinsey and Company, March 2020).
8. UK Ministry of Defence, *Global Strategic Trends: The Future Starts Today*, 6th ed. (Swindon: Development, Concepts and Doctrine Centre, 2018).
9. National Intelligence Council, *Global Trends: Paradox of Progress* (Washington, DC: National Intelligence Council, January 2017).
10. Government of the Republic of France, *Defence and National Security Review 2017* (Paris: Government of the Republic of France, 2017).
11. Russian Federation, *National Security Strategy* (Moscow: Ministry of Foreign Affairs, 31 December 2015).
12. State Council Information Office, *China's National Defense in the New Era* (Beijing: Chinese Communist Party, 24 July 2019).
13. Daniel R. Coats, *Statement for the Record: Worldwide Threat Assessment of the U.S. Intelligence Community* (Washington, DC: Office of the Director of National Intelligence, 29 January 2019).
14. Roman Muzalevsky, *Strategic Landscape 2050: Preparing the U.S. Military for New Era Dynamics* (Carlisle, PA: Strategic Studies Institute and U.S. Army War College Press, 2017).

Fukuyama subsequently doubled down on his argument with his 1992 book, *The End of History and the Last Man*.[42] In the book, Fukuyama argues the evidence for the core idea described in his *National Interest* article. He notes that "while the book is informed by recent world events, its subject returns to a very old question: whether, at the end of the twentieth century, it makes sense for us once again to speak of a coherent and directional history of mankind that will eventually lead the greater part of humanity to liberal democracy. The answer I arrive at is yes."[43]

The intervening time has shown that there is no global consensus about a single form of government. Contrary to the hopes of Fukuyama, events have not moved in that direction. As an article in *The Atlantic* noted in 2014, "There's a growing sense that liberalism isn't delivering at home and that it's not as popular as we think it ought to be in the developing world."[44] Surveys of democratic systems point to a retrenchment from liberal democracy.[45] In a 2015 article in the *Journal of Democracy*, Larry Diamond described this as a *democratic recession*.[46]

The level of competition has already led scholars such as Robert Kagan to declare that "the liberal world order established in the aftermath of World War II may be coming to an end."[47] Whether or not this is the case, it is an environment that we must understand in coming to grips with change in the twenty-first century. Understanding this changed environment, and the changes and instability it will drive in the policy and strategic settings of different nations, will offer an important context for the curricula of all military educational institutions.

The new era in geopolitics has many influences. However, among the threats from nonstate actors, nuclear-armed dictatorships such as North Korea, and threats to the legitimacy of democratic governance, there are two principal drivers of change in geopolitics: China and Russia. A reemerging China and a reassertive Russia are forcing evolution in the character of international politics, trade and economics, diplomacy, and military operations. This is happening more rapidly than at any time since the end of the Cold War. As a 2018 RAND report notes, this new environment is likely to be characterized by "intense competition between a handful of states" and will feature competition "between the architect of the rules-based order and the leading revisionist peer competitor."[48]

Russia and China have their own peculiar views of the world. They seek to dominate their immediate regions, while also shaping global economics and politics to align with their interests.[49] Each nation has invested heavily in conventional military capabilities. For China, the ongoing investment in conventional forces is part of the Chinese Communist Party's aim of exploiting a period of strategic opportunity, building a "world-class military," and becoming a great power.[50] Russia is challenging the current status quo, partially through its reinvigoration and reinvention of strategic art, applying all of its tools of state power in an integrated, single whole.[51]

Russia and China are also engaged in their own forms of influence and political warfare. They do this to manipulate public and domestic political debate in Western nations so they might increase their influence while undermining that of democratic nations.[52] Russian general and theorist Valery Gerasimov has described the Russian strategy for dealing with the United States as preemptively neutralizing threats, formulating strategies of limited action, and understanding how "military science needs to develop and substantiate a system for comprehensive engagement of the enemy."[53]

The challenge posed by China is of a different magnitude. Bruce Jones describes China and the United States as "the central factor in every other actor's calculation."[54] The Chinese regime strives to achieve three primary goals: to maintain undisputed Communist Party rule, to continue the growth of the national economy in order to develop a wealthy nation that is united and loyal to the regime, and to enhance China's international influence and prestige to be an equal, if not superior, rival to the United States. Over the past decade, China has made substantial advances in its regional and global influence through its persistent political warfare operations.[55] It is using four key levers of national power to achieve its aspirations: political warfare, economic power, technology development, and military power.

Oriana Skylar Mastro has written that "throughout history, would-be powers have invented new ways of growing."[56] In this vein, China has sought new sources of power. It has conducted overt and covert political activities to

neutralize or even co-opt opposition from foreign powers. In a 2019 examination of this phenomenon, Ross Babbage explored multiple case studies of
the Chinese government using political warfare to gain leverage in the Indo-
Pacific region while seeking to diminish that of its principal competitor, the
United States. Babbage highlights the importance of Western democracies
understanding this dimension of competition when he notes that "political
warfare operations have been central to the Chinese and Russian regimes'
strategic advances for the last two decades. Both regimes are well equipped,
very experienced, and highly skilled in the conduct of these political warfare campaigns. The West, by contrast, largely abandoned high-level political warfare operations at the end of the Cold War."[57]

In the past decade, Russia has been conducting its war in Crimea and
Ukraine with a form of political warfare called *reflexive control*. The Russians
have been skillful in the use of this technique, particularly in convincing
Western governments to do something they were already inclined to do:
remain on the sidelines of the Russian operations in Ukraine. Reflexive control has relied on denial and deception to mask the presence of Russian
forces, to obscure Moscow's strategic goals, to maintain superficial plausible
deniability, and to threaten Western governments.[58]

China also possesses a sophisticated capacity for political warfare.
Constructed on deep historical preferences for indirect forms of achieving
its strategic goals, the Chinese Communist Party is well equipped, highly
motivated, and very skilled in the conduct of this aspect of strategic competition. Their skill has contributed to a range of successful outcomes for the
Chinese over recent years, including their emergence as a dominant power
across the Indo-Pacific region and their effective seizure and militarization of
the South China Sea.[59]

The Chinese Communist Party has been inventive in the use of economic
power to attain its strategic objectives. Probably the best-known example of
this is its Belt and Road Initiative, a program launched in 2013 that comprises a massive regional infrastructure program. Some analysts see the project as an unwelcome extension of China's increasing global heft, particularly
given that it could use regional development as a Trojan horse for expanding its military footprint.[60] Having spent nearly half a trillion dollars on this

initiative, the Chinese so far have convinced eighty-six nations and international organizations to join what Chinese president Xi Jinping describes as a "community of common destiny."[61]

China is using the wealth it has generated over the past two decades to invest in technological and social development as well as to grow its military power and global influence. These goals were reflected in Xi Jinping's three-hour speech at the nineteenth Communist Party congress in October 2017. He described China as a great power twenty-six times and committed to building a more powerful military institution that would be "first class in every way" and ready to fight.

The technological advances China has made in the last two decades have been stunning by any measure. It has advanced its ability to research, develop, and deploy new technologies across the fields of information and biotechnology, energy, and new materials. While China has benefitted from what some have called the largest theft of intellectual property in history, Beijing has increasingly developed its indigenous research and development capacity. The efforts of the United States and other nations to address the theft of intellectual capacity have reinforced the Chinese government's efforts to push harder to achieve technological self-reliance.[62]

In 2015 China announced its *Made in China 2025* strategy, which described its aspiration to lead the world in high-tech industries such as aerospace, information technologies, robotics, and other sectors.[63] It seeks to end its reliance on international technology and to upgrade its capability across ten key industries.[64] China has matched this intent with massive funding of research and development, as well as the creation of new innovation centers. The strategy aspires to move China up the value chain through smart manufacturing and to become a global leader in a variety of innovative industries by 2025.

Like all empires through history, however, the emergent Chinese empire has weaknesses. Internal weaknesses include domestic support, an aging population, economics, pollution, and governance. At the same time, China must balance investment in maintaining a large ground force to protect its borders, building a large naval force that can project power beyond the first island chain in the Western Pacific, and sustaining the capacity to threaten (or eventually seize) Taiwan.

Several authors have examined domestic and strategic Chinese weaknesses in the last two decades. William Overholt has observed that "China seeks to cope with its crisis of success; it has tremendous strengths and tremendous weaknesses."[65] Overholt proposes that China is facing multiple political risks as it seeks to transition its economy and overcome social challenges that include inequality and environmental degradation. Susan Shirk finds that "China's growing economic might makes its leaders appear to loom over us like ten-feet-tall giants. But their self-image differs vastly. Dogged by the specters of Mao Zedong and Deng Xiaoping . . . China's current leaders feel like midgets, struggling desperately to stay on top of a society roiled by economic change. China may be an emerging superpower, but it is a fragile one."[66]

A 2015 RAND Corporation study found that despite significant advances in Chinese military capabilities since 2000, important vulnerabilities remained. These include the ability to undertake joint operations, training and education of personnel, integration of complex technical capabilities, and rampant corruption. Understanding these military weaknesses provides insights into potential future areas of military reform in China, as well as indicating areas that might be leveraged in a new approach to deterrence.[67]

### Second-Tier Competitions

The U.S.-China strategic competition will dominate geopolitics in the coming decades, but the grand strategic competition between these countries is not the sole story. A range of subsidiary competitions will influence the global security environment while also driving the national security policies of the United States, China, and their allies and trading partners. While each might share some of the common elements already explored, the context and strategic cultures of participants in each can differ markedly.

The Pakistan-India competition is critical to understand because it involves two nations with a combined population of more than 1.5 billion people,[68] and both possess large stockpiles of nuclear weapons. The India-China competition is important because each represents a very different political ideology and has large nuclear and conventional forces, and a large influx of Chinese capital and military power into the Indian Ocean region has occurred. Finally, the U.S.-Russia competition, while largely decided at the conclusion of the Cold War, remains an important aspect of geopolitics because of the respective sizes of their nuclear arsenals and Russia's desires to maintain a large sphere of influence in its west and to play the part of a strategic spoiler for the exemplar of democratic systems.

This second level of competition and cooperation that exists beneath the superpower competitions is an important aspect of the regional and global order. But so too is the impact of multinational institutions, which remain part of the global system. Despite their poor showing during the COVID-19 pandemic and the impact on the legitimacy of institutions such as the World Health Organization, multinational institutions remain a persistent element in the security environment. Other institutions such as the United Nations retain enormous influence in large parts of the world and provide a source of some power for small and mid-size countries. Some will continue to push for reform in the United Nations and to have a larger group of permanent representatives in its security council, but it will remain a relevant and influential actor in the global security environment in any event.

The U.S.-China competition, the subsidiary competitions between lesser powers, and the impact of multinational institutions create a complex and confusing environment. Interactions occur constantly, and tracking the multitude of different and evolving relationships between nations, regional blocks, and nonstate entities is impossible. The impact of the recent pandemic will play out over many years. However, as historical pandemics impacted existing orders, so too this most recent one will potentially hasten some aspects of change and disruption in the current system.

A multipolar global system has returned. As a 2016 U.S. Joint Staff publication described, the future environment will see "a number of states with the political will, economic capacity, and military capabilities to compel change at the expense of others." After nearly two decades of struggle against violent extremists and operations in the Middle East, the United States and its allies are beginning to adapt to a world in which great power competition is once again a dominant factor in national security planning.[69]

And while history, particularly Cold War history, provides a guide, ideas such as containment are unlikely to provide the right intellectual foundation for the coming decades of strategic competition.[70] Similarly, the twentieth-century approach to deterrence, based on holdings of large quantities of nuclear weapons, will need to be reexamined in light of a competitor that prefers competition beneath the threshold of violent conflict. This will demand evolved thinking about how nations, large and small, might overcome coercion by large nations and how nonnuclear elements of national power might be used to deter threats of violent conflict.

In this new era of strategic competition, two great powers, the United States and China, coexist in a new form of "cold but connected war." As Alan Dupont has written, "Technology and geopolitical conflicts have precipitated a new Cold War. In an epoch-defining clash for global leadership, the world's two major powers are wrestling for strategic advantage in an increasingly bitter contest to determine which of them will be the pre-eminent state of the twenty-first century."[71] Every nation and every military institution in the world will be affected by this competition.

## DISRUPTOR TWO: DEMOGRAPHIC SHIFTS

Demography, the study of statistics such as deaths, marriages, births, and education, allows a better understanding of the changing nature of human populations. For governments, demographic knowledge underpins their delivery of services to the populace. For military leaders and national security practitioners, the study of demography can provide vital insights into the capacity of nations to recruit and sustain military forces as well as the types of human environments and conflicts that they may have to prepare for in the future.

Demographic issues have challenged governments and scientists for centuries. Thomas Malthus, concerned that humans would outgrow the resources available to them, published *An Essay on the Principle of Population* in 1798.[72] It was the first book to consider population, or demographic change, and it contributed to the establishment of a national census in the United Kingdom in 1801.

Malthus was principally concerned about what the number of humans populating the planet was that would be beyond the capacity to feed. He proposed models to estimate future population growth, writing that "in two centuries and a quarter, the population would be to the means of subsistence as 512 to 10,"[73] as well as how society might limit growth (such as lowering birth rates). But his work is best known as the origin of the term *Malthusian trap*—a situation where a large population would cease growing due to food shortages, which would then lead to starvation.

Despite a global population that has grown from 1 billion to nearly 8 billion since Malthus published his first edition, predictions of calamity have not eventuated. A steady increase in global population has moved in step with increases in technology and agricultural productivity and with demographic shifts.

The world's population continues to rise, standing at 7.9 billion people in April 2021. The United Nations projects the world population growing to 10.9 billion people by the end of this century.[74] However, some have painted a much different picture for global population prospects. Using an evolved analytical model that incorporates educational outcomes, the Austrian International Institute for Applied Systems Analysis projects population to peak at 9.7 billion before falling to between 7.8 billion and 9.3 billion by 2100.[75]

The United Nations projections for growth in the global population through 2100 are also disputed by Darryl Bricker and John Ibbitson, who propose that "the great defining event of the twenty-first century—one of the great defining events in human history—will occur in three decades, give or take, when the global population starts to decline. Once that decline begins, it will never end. We do not face the challenge of a population bomb but of a population bust—a relentless, generation-after-generation culling of the human herd. Nothing like this has ever happened before."[76]

Bricker and Ibbitson base their argument on declining fertility rates and the unlikelihood of their recovery over the course of the twenty-first century. Urbanization is a key part of this story. As more people move to cities, they are having smaller families because it is more expensive to raise children in cities. More importantly, urban women have better access to education and more opportunities to congregate with other women, which results in greater autonomy and more control over the number of children in a family. Finally, this decline in the number of people may also have significant implications for national security. As the authors note, "Population decline will shape the nature of war and peace in the decades ahead, as some nations grapple with the fallout of their shrinking, aging societies while others remain able to sustain themselves."[77]

Whether the global population continues to grow, stabilizes, or starts to decline over the coming decades, knowledge of demographic trends will have an influence on how nations plan for their future national security. There are several demographic trends that impact government planning, national security strategies, and the future of military power.

One important demographic trend occurring around the world is population aging. As a recent United Nations report finds, "Population aging is a human success story, reflecting the advancement of public health, medicine, and economic and social development, and their contribution to the control of disease, prevention of injury, and reduction in the risk of premature death."[78] Almost every nation in the world is experiencing growth in the number and proportion of older people in their population.

Since 1950 the life expectancy at birth for the world has increased by 28 years, from 45 to 73. In many nations, the average life expectancy is more than 80 years. Globally, the number of older people (those aged 65 and over) is growing more quickly than the number of people in any younger age group. They now represent nearly 10 percent of the world's population, and the United Nations expects there to be 1.5 billion people over the age of 65 by the year 2050.[79]

Aging is not the only demographic challenge for national leaders and those responsible for national security policies. Another important demographic trend is urbanization. The percentage of the world's population residing in urban areas has grown significantly over the past 70 years—from 751 million in 1950 to 4.2 billion in 2018. In 2018, 55 percent of the world's population lived in urban areas. Migration flows have also changed over the last several decades. From 1990 to 2017, the number of international migrants rose by 69 percent. For some nations, regulated migration flows ensure an inflow of younger people and capital, which enrich the recipient nation. In other nations, unregulated flows have resulted in inequalities and societal tensions.[80]

Urban areas are important drivers of economic, cultural, and social progress. But the increase in urbanization will create challenges as well. As people urbanize, they also put pressure on governments for more services. They become better educated but also have higher living standards, which has historically resulted in lower birth rates. As a result, we are currently seeing the birth rates in almost every nation in the world decline. Humans face the potential of an accelerating demographic shock where the population of nations including China, Japan, Russia, and all of Europe declines. In addition, many large cities are coastal and thus are particularly at risk of rising sea levels due to climate change.

A final demographic trend is the large-scale migration of people in different regions and across the globe. People leave their home regions or nations for a variety of reasons. Most migrate internationally for reasons related to work, family, and study. Around half of migrants move from developing to developed nations, where immigration is a key contributor to population growth. These immigrants have contributed up to 80 percent of labor force growth in the past two decades in destination countries. In a 2016 investigation, the McKinsey Global Institute found that migrants contributed approximately $6.7 trillion (or around 9 percent) of the global gross domestic product in 2015.[81]

Other people leave their homes and countries for more compelling reasons, which can include persecution, natural disasters, environmental degradation, and conflict. While these refugees and internally displaced persons comprise less than 10 percent of the global migrant population, they are often the most in need of aid and support.[82]

## DISRUPTOR THREE: NEW AND EVOLVED TECHNOLOGY

The search by humans for the ultimate tool to influence, defeat, or destroy their adversaries is a story almost as old as *Homo sapiens* themselves. For millennia, human beings have adapted existing tools or developed new technologies to produce better weapons. For the last two centuries, countries in the West have relied on some form of technological edge to sustain a relative advantage over conventional and unconventional threats.[83]

New technologies have fundamentally shaped and reshaped military institutions, and their operating theories, for almost as long as there have been military organizations. Williamson Murray and MacGregor Knox have proposed that the first major disruption related to technology occurred in seventeenth-century France. The flintlock fusil with bayonet and cannons were combined with new drills, a battle culture of forbearance, and modern military institutions. This resulted in new approaches to war that shaped all future development, as well as underpinning waves of European colonial expansion.[84] This will remain the case in the century ahead of us.

Eight specific technologies—artificial intelligence, robotics, quantum technology, biotechnology, energy weapons, hypersonics, space technology, and additive manufacturing—are likely to have the greatest impact on twenty-first-century military affairs. These technologies will mainly (but not exclusively) impact society first and then flow into military institutions.

### Artificial Intelligence

Artificial intelligence (AI) involves the application of computers to simulate activities that have traditionally required human intelligence (or human cognition). The *Bulletin of the Atomic Scientists* has noted that AI "is something a computer or a machine does upon interacting with the environment to achieve certain goals by means that mimic human cognitive functions."[85] The continued improvement in computing performance has been a fundamental foundation for advances in AI over the past two decades.

Most contemporary artificial intelligence is more accurately described as *narrow AI*, which performs narrowly defined tasks (for example, facial recognition or only Internet searches or only driving a car). Streaming devices on televisions and mobile devices are forms of narrow AI. The capacity to operate at the level of human intelligence in broader or nonprogrammed tasks is called *artificial general intelligence* (AGI). While narrow AI may outperform humans at whatever its specific task is, like playing chess or solving equations, AGI would outperform humans at nearly every cognitive task.[86]

There are a range of functions where AI may extend cognition and permit the development of an AI-enhanced human intellectual edge. For example, AI has wide application in large-scale information analysis. It will continue to underpin a large variety of autonomous systems on the ground, in the air, and on (and below) the sea. It will be a powerful tool to support research and development across a range of technologies with military applications including hypersonics, quantum computing, aerodynamics, hydrodynamics, ballistics, material sciences, and biotechnology.

At the most basic level, AI will change the balance of power in tactical military endeavors. Kenneth Payne proposes that this "will change the utility of force by enhancing lethality and reducing risk to societies possessing AI-warfighting systems. . . . A marginal technological advantage in AI is likely to have a disproportionate effect on the battlefield."[87] It offers multiple possibilities for decision support at the tactical level of war. A capacity to support alignment of tactical goals with higher aims is just the tip of the iceberg. Using human-in-the-loop and human-on-the-loop systems, forms of AI may be applied for rapid decision-making. Other forms of AI might support the integration of joint capabilities and assist in tactical planning through rapid simulation of the outcomes of multiple options before and during military activities.

Artificial intelligence has ramifications for strategy development as well. Colin Gray has written that "the most enduring function of strategy is management of potentially lethal dangers. Strategists need to be 'right enough' to enable us to survive the perils of today, ready—and possibly able—to cope strategically with the crises of tomorrow."[88] James Holmes and Toshi Yoshihara have described how "the realm of ideas is admittedly more nebulous than that of net assessment, but it is just as consequential. Understanding how an aspiring or established power thinks about strategy and may try to put strategic ideas into practice is indispensable to forecasting that contestant's fortunes."[89]

But the current pace of technological change is disrupting how nations and their military institutions understand other nations and develop the various strategies that are required to achieve their desired national objectives. Human-only approaches to strategic decision-making may become insufficiently effective against adversaries that will probably apply AI to model, test, and execute military activities at the strategic level. In the coming years, it is almost certain that humans will require AI to assist them to develop, test, and implement strategies related to defense and national security.

The ethical challenges inherent in using machines to kill without human input have been discussed in a range of publications over the last decade.

Humans making the final decision with lethal systems represent what Paul Scharre has described as a "bulwark against malfunctions and flawed interpretations of data."[90] At the same time, the push for military leaders to use AI-based predictive analytics has its own challenges. As Heather Roff notes, "Predictive models of social phenomena can at best rely on crudely coded past events, where those past events are shaped by a myriad of latent processes and variables unaccounted for by the relevant data. Furthermore, AI systems are not value-free and neutral. Rather, they are inherently biased because of their subjectivity. . . . We should be careful not to make assumptions about the capabilities of prediction machines and should seek to understand how they work, when they do not, and why."[91]

Ethical challenges will pose obstacles for those developing the various algorithms to support human decision-making. However, this will not slow down the push for military institutions at every level to use AI for more functions. One of the greatest challenges of the coming years will be how we might achieve an optimal balance of technological capacity in the use of artificial intelligence with an ethical approach to decision-making.

## Military Robotics

While military robots have existed since World War II, it was the post-9/11 wars that kick-started a modern "robot rush." With permissive airspace and very little threat in the electromagnetic spectrum, the U.S. military and its partners poured investment and procurement dollars into persistent aerial surveillance systems that are now pervasive in operational environments. Not long after these new surveillance drones were deployed in places such as Iraq and Afghanistan, a range of other military uses were discovered for uncrewed robotic systems. These included disposing of improvised explosive devices, searching dangerous or restricted areas on or under the ground, and logistic support.

Robots, and robots in human-robot teams, have a variety of future military applications. Human-robot combinations will be useful in training establishments to improve training outcomes and to provide the test-bed for best practices in developing human-robot tasking relationships. In logistics, robots will have utility in performing tasks that are dirty, dangerous, or repetitive—in contaminated areas; urban, deep sea, and subterranean environments; densely protected military sites—but also in performing more mundane tasks such as vehicle maintenance and repair and basic logistics and movement tasks.[92]

The use of human-robot teams during operations offers a solution to an enduring challenge for many military institutions: building mass and doing it in different locations concurrently. Potentially, individual humans might control a small fleet of ground and air systems, providing for an exponential increase in the capability of deployed forces. A 2016 U.S. Department of Defense experiment saw more than one hundred micro drones released to form an autonomous swarm,[93] and a 2021 U.S. Navy experiment tested ways to integrate crewed and uncrewed naval vessels and use "super swarms" of autonomous systems.[94] The Chinese have conducted multiple experiments with swarming killer drones.[95] As swarming concepts are more widely employed, future military institutions will have to invest in a range of "mothership" functions across the land, sea, and air domains.[96]

A highly capable joint military task force in the future might consist of as few as several hundred human personnel and several thousand robotic systems of various sizes and functions. Many functions currently undertaken by humans might be better performed by robots in human-robot teams; this has the potential to reduce the size of many types of military units by hundreds of personnel. The U.S. Army even described in detail such an approach in a 2020 presentation called "Robotic Warfare Battlefield Geometry."[97] Concurrently, the Chinese PLA has introduced into service its armed Sharp Claw autonomous ground vehicle.[98] The ongoing introduction of these systems will free up personnel for redeployment into areas where the art of war demands leadership and creativity—enabling intelligence functions, training and education, planning, and, most importantly, command and leadership.[99]

As with AI, there are ethical challenges with the use of military robots. Who decides which machines kill, and when? What limits might be placed on the killing power of autonomous systems in the air, on land, and at sea? Are drone swarms actually weapons of mass destruction?[100] These questions remain to be answered satisfactorily by governments or contemporary military institutions. But there are limited prospects for slowing the development and deployment of lethal autonomous robotic systems. While previous generations of national leaders collaborated to regulate weapons such as poison gas and nuclear weapons, this has not been the case with military robotic systems. The major powers—the United States, Russia, and China—have all obstructed efforts to ban such systems.[101] For now, we should expect to see a continuing proliferation of these systems in the military organizations of nation-states and, increasingly, of nonstate actors.

## Quantum Technology

Quantum technology is a relatively new field of engineering and physics that harnesses the strange and often misunderstood properties of quantum physics. All of the potential applications of this technology use several fundamental properties of quantum phenomena. The concept of superposition refers to the ability of a particle to exist across all possible states at the same time. Another concept, entanglement, involves linkage among two or more particles such that their properties are interrelated. Concepts such as tunnelling, quantization, and decoherence add to the strange properties of quantum mechanics and give them their unique power and potential.[102]

In 2017 China announced a project to construct the first "unhackable" computer network using quantum technologies. This built on the 2016 launch of a quantum satellite by the Chinese. Other nations have spent decades investing in quantum technology to achieve more cost-effective and secure movement of data.[103] They have done so because quantum technologies have a range of future theoretical applications including secure communications, sensing, and computation.[104]

Quantum computing relies on quantum bits (qubits) that can exist in superimpositions of 1 and 0, meaning they can be both 1 and 0 at the same time. The result is a massive, even exponential, increase in processing and information capacity. These computers hold the promise of solving problems beyond even the most capable supercomputers. To maximize their potential, the quantum computers will also require new types of algorithms (quantum algorithms). Quantum technology is likely to be best suited to problems that possess multidimensional parameters and that require the optimization of large numbers of variables. These might include breakthroughs in materials science, new approaches to machine learning for AI, as well as improvements in military logistics and cracking resistant encryption.[105]

Quantum cryptography and communications also have a variety of potential applications. China is investing heavily in deploying quantum communications networks to secure its most sensitive communications. Its established national quantum network is experimenting with quantum encryption. If this proves to be a scalable approach, China conceivably could roll out a national, secure quantum communications network over the coming decades.[106] Not only will this pose challenges for those wishing to intercept military and government command and control networks, but it also may make the secure communications networks of every other nation obsolete.

Quantum timing and navigation are other potential applications with both civilian and military uses. The employment of quantum phenomena may underpin the development of quantum "clocks" that could provide hyper-accurate timing. This would have application in areas such as algorithmic trading in the commercial world and highly precise target location systems in the military world. Such a technology might improve underwater navigation by submarines, improve the precision of weapons systems, and also serve as a useful redundant navigation system should global positioning systems (GPS) be denied by enemy jamming or spoofing.[107]

Other theoretical applications of this technology include improving situational awareness through different sensors, improving accuracy of sensors, and offering new materials for use in semiconductors. However, quantum technology remains experimental in many respects. Despite the massive investments in this technology in different nations and the race to develop quantum computers, functions such as reliable quantum cryptography remain unlikely before 2030. However, when such devices do become available and begin to proliferate through government, commercial, scientific, and military institutions, they will revolutionize computing.[108]

### Biotechnology

In the 1970s U.S. scientists at Stanford University proposed the concept of recombinant DNA, which has permitted biology to move from an exclusively analytical science to a synthetic one. In the decades since, biotechnology has advanced as rapidly as information technologies have but without a similar level of attention.[109]

Revolutionary breakthroughs have occurred in this field of science over the past decade. Human genome sequencing has progressed from a multiyear and multi-billion-dollar undertaking to one that is much cheaper, quicker, and more accessible. The reductions in cost in genome sequencing since 2000 have far outstripped Moore's Law.[110]

The speed and reliability of gene editing now achievable with technologies such as the CRISPR/Cas9 system are affecting fields from basic research to therapy. Berkeley biochemist Jennifer Doudna transformed CRISPR/Cas 9 from a bacterial curiosity into a tool that can edit genes in any organism. The technology has resulted in advances in agriculture, medicine, environmental science, and other endeavors. Previously, genetic engineering was time-consuming, inexact, and expensive. This all changed with this revolutionary gene-editing technology.[111]

The potential applications of this technology are numerous. For example, trials have been undertaken using this technology on cancer patients. Patients in U.S. medical trials have been injected with immune cells gene-edited using CRISPR.[112] It has also been used for developing novel materials that can change their properties when they are exposed to different DNA sequences. This has application in controlling electronic circuits and microfluidic devices as well as diagnostic activities or delivering medical treatments.[113] Military applications could include increasing human strength and endurance, rapid repair or replacement of damaged organs and tissue, or even improving human cognitive skills.

These advances in biotechnology will fundamentally change how military institutions recruit, develop, and exploit their future human capacity. But this new age of biotechnical capability must be matched by enhanced human responsibility. In many respects, the ethical considerations of biotechnology mirror those of artificial intelligence and lethal autonomous robotics. It is less a question of whether these technologies *can* be developed than whether they *should* be developed. Governments and military institutions will need to ensure that developments in ethical and lawful use of biotechnologies keep pace with technological progress.

## Energy Weapons

Military institutions have long desired the ability to leverage energy for destructive purposes. Historians are still debating whether Archimedes built a weapon that used bronze or copper shields as mirrors to focus sunlight to destroy Athenian ships, as described in the writings of second-century AD author Lucian. While contemporary military forces will not be using mirrors to focus the rays of the sun on their adversaries, directed energy weapons remain an attractive proposition.

The U.S. Defense Advanced Research Projects Agency (DARPA) has had several programs over the past decade to explore this type of weapon, including its high-energy liquid laser area defense system.[114] Directed energy systems offer the potential of more effective future defensive systems against intercontinental ballistic missiles, other rockets, mortars, and artillery. As well as lethal effects, directed energy can be used for nonlethal coercive applications.

In recent years, research has focused on development of chemical-based and electric laser weapons. The U.S. Army has awarded contracts to Raytheon and Northrop Grumman for the development of ground-based laser weapons to disable rockets, artillery, and vehicles.[115] Companies such as Lockheed Martin are also developing laser weapons for a variety of land, air, and sea missions.[116] The U.S. Navy plans to deploy its high-energy laser and integrated optical-dazzler and surveillance system on a destroyer in 2021.[117]

Nations such as Russia are investing in these weapons as well. In 2017 Russian news agency Pravda reported that the Russian air force would arm its aircraft with laser weapons.[118] China is also investing in laser weapons, with one weapon in the 50- to 70-kilowatt range, mainly designed for antidrone weapons.[119] It also has a long-standing program at the Chinese Academy of Sciences[120] to develop ground-based lasers for counterspace missions as well as for satellite laser ranging applications.[121] Even Great Britain is researching laser weapons, unveiling its Dragonfire laser turret in 2017.[122]

Whether they are chemical or electrical in nature, these systems suffer from a range of problems include reliable, deployable sources of energy and the lack of maturity of the laser technologies. However, at some point, these hurdles will be overcome. And while soldiers are unlikely to be carrying lethal laser weapons in the near future, these weapons will fill a series of niche defensive applications in operations at some point in the twenty-first century.

## Hypersonics

Hypersonic vehicles are those that fly at speeds of at least Mach 5 (five times the speed of sound) through the atmosphere (not space). Humans first achieved hypersonic speeds in the late 1940s when the U.S. two-stage Bumper rocket reached speeds of 8,288 kilometers per hour (around Mach 6.7) in tests at White Sands Missile Base. The first human to travel at hypersonic speed was Russian Yuri Gagarin during his historic orbital flight in April 1961.

Over the past four decades, nations such as the United States, China, Russia, France, the United Kingdom, Japan, India, Australia, and others have invested in the research that underpins hypersonic flight. A 2017 RAND report surveyed the efforts of more than twenty nations (as well as the European Union) in developing hypersonic capabilities.[123] They have done so because this technology has application for air transport. However, while transportation is an important area of research, it is secondary to the current focus of research: high-speed weapons. Hypersonic weapons have three advantages: reduced detection, compressed decision times for defenders, and higher destruction from high speed.

Detecting and tracking hypersonic weapons are much more technically difficult than doing so for other conventional missiles, rockets, and aircraft. Space-based and terrestrial sensor systems are currently insufficient for this particular mission. Hypersonic glide weapons generally coast toward their targets at altitudes of around 30 to 50 kilometers above the ground. This is much lower than traditional ballistic missiles, which are fired into outer space beyond the 100-kilometer Kármán line. Ground-based systems will have less time to detect them, and more investment in air- and space-based sensors will be required for useful defensive systems.[124] As U.S. Secretary of the Army Ryan McCarthy noted in a 2020 presentation, the threat posed by these weapons will demand wider arrays, a low-Earth-orbit satellite constellation, and a better capacity to queue targets rapidly.[125]

Hypersonic weapons not only compress the time available to decision-makers whose military systems are targeted (ensuring that automated interception may become necessary) but also limit the number of times that defensive systems may be able to attack incoming hypersonic weapons. These weapons are particularly effective in tactical situations where ranges might not be intercontinental but rather in the hundreds or low thousands of kilometers. This makes them more likely to be able to penetrate existing missile defense systems. A 2017 RAND report notes that "the trajectory and capabilities of these weapons provide them with some unprecedented attributes that may be disruptive to current military doctrines of advanced nations. They can defeat current ballistic missile defense systems because of their unpredictable long-range trajectories, maneuverability, and flight altitudes. Hypersonic weapons substantially increase the threat for nations with otherwise effective missile defenses."[126]

Finally, an attribute that is common to all forms of hypersonic weapons is that high-speed energy allows them to solely use kinetic energy to damage or destroy targets. In combining high accuracy and high speed, a range of different targets, including those that might be buried underground, becomes more vulnerable. For example, a hypersonic missile with a 500-kilogram kinetic warhead (having no explosive content) impacting a target at a speed of Mach 8 possesses the equivalent of around three tons of TNT. This introduces a new form of vulnerability for static targets and those that are hardened in various ways, including siting them underground.[127]

### Space Technology

The first space age, from the 1950s through the 1990s, witnessed the development of a range of technologies, such as rockets, GPS, and satellites. It also produced training systems and the scientific knowledge for humans to journey beyond the edge of the Earth's atmosphere. As Carl Sagan noted in 1987, this "time will be remembered because this was when we first set sail for other worlds."[128] Despite these great advances, space travel and the launch of satellites into orbit were still an expensive and time-consuming endeavor.

The economics of the space industry started to change in the 1990s. With the collapse of the Soviet Union, Soviet rockets and other technologies were offered on the world's private markets. Assisted by U.S. firms, new companies such as International Launch Services and Commercial Space Management paved the way for private industry's utilization of space. In the early 2000s, entrepreneurs in the United States (mainly from Silicon Valley) started pouring private investment funds into space flight activities.

The magnitude of the resulting cost reduction, as well as advances in smaller and cheaper satellites, is driving what some call Space 2.0. This is resulting in an accelerating pace of satellite launches with companies such as SpaceX launching hundreds of satellites to provide global Internet connectivity. Its StarLink system, launching 60 satellites at a time on its Falcon 9 rockets, is intended to eventually incorporate a constellation of over 12,000 satellites orbiting at 550 kilometers above Earth.[129]

A new competition is brewing between the United States, China, and Russia. China and the United States are both investing billions of dollars in a new race to return to the moon. A return to the moon and cheaper space launch capabilities are just two dimensions of the new space age. China and Russia are also developing jammers, lasers, and cyberattack and kinetic kill capabilities for use in space. This approach is hardly new. During the Cold War, both the United States and the Soviet Union pursued technologies to destroy or compromise the satellite capabilities of their adversaries with direct ascent or co-orbital systems.

In the twenty-first century, the United States, Russia, China, and India have all tested direct ascent antisatellite missile capabilities. The Chinese conducted their first test in 2007, destroying one its own defunct satellites. The Indians conducted a successful test in March 2019. However, these capabilities are expensive and complex to develop and maintain. Because of this, several nations are developing another method of interfering with, or destroying, satellites: co-orbital weapons.[130]

Co-orbital weapons are orbiting space craft that maneuver and intercept satellites in orbit. This interception may be intended to interfere with the purpose of the targeted satellite, to disable it, or to gain intelligence on its capabilities. During the Cold War, Soviet antisatellite capabilities focused on co-orbital systems. Launching its first generation of these in 1963 called *Polyot-1*, the system entered service in 1978 as an interceptor launched atop an R-36 intercontinental ballistic missile. The Soviets called their co-orbital system *Istrebitel Sputnikov* (fighter satellite). But at the end of the Cold War, the Boris Yeltsin government decommissioned this capability.[131]

Space-based systems are a central capability in the global economy, national communications, and military operations. They provide the ability to communicate and to collect imagery (important for many applications beyond military intelligence), and they provide precision navigation and timing information used by every industry and individual with a personal communication device. We should expect the competition for cheap and assured access, as well the development of antispace capabilities, to continue and accelerate over the coming decades. It is a capability that symbolizes the true high ground of twenty-first-century competition and conflict.

### *Additive Manufacturing*

In May 1980 Hideo Kodama of the Nagoya Municipal Industrial Research Institute in Japan filed the first patent related to three-dimensional (3D) printing. The patent described a process that used photopolymer material, exposed to ultraviolet light, for use in rapid prototyping. However, this process was never commercialized. Not until 1986 did the 3D Systems Corporation develop the first commercial 3D printing system, the SLA-1, which used a method that printed objects layer by layer using lasers. The process has come to be known as 3D printing or additive manufacturing.

By 2020 the additive manufacturing industry encompassed aerospace, automobile manufacturing, medicine, and education. From 2014 to 2019, corporations such as General Electric, Google, Norsk Titanium, and others were investing hundreds of millions of dollars in the development of this technology.[132] New versions of 3D printing technology are expected to continue advancing in capability and affordability in the coming decades. This will allow the use of new materials, increase production speeds, and lower manufacturing costs. Given the advantages of industrial 3D printing, it is increasingly becoming part of many nations' manufacturing capability. It is also a technology that appears to offer interesting new applications in the realm of military endeavors. Some of the earliest investors in 3D printing research were military institutions. Both the U.S. Office of Naval Research and DARPA provided "steady, continual streams of funding for both academic and industry-based researchers."[133] The military's interest in additive manufacturing continues unabated.

T. X. Hammes has written extensively on future warfare and new technologies. He has examined many different technologies for their impact on war and has described the implications of additive manufacturing for military operations as "immense."[134] As a technology that reduces the cost of building complex items and enables it to be done in dispersed locations, 3D printing is very attractive to military organizations. Researchers at the University of Virginia printed an unmanned aerial vehicle in a single day at a cost of $2,500. Companies such as United Parcel Service offer large-scale 3D printing capacity at multiple locations throughout the continental United States,[135] which offers lessons on the distributed use of this technology by military organizations.

Just as Henry Ford probably did not conceive of the many applications of the moving assembly line he invented, so too we have not yet fully explored the possibilities of 3D printing. It has significant potential in civil industry, particularly for improved productivity and more rapid product development cycles. Its potential for improving logistics and reducing the cost of mass-produced complex machines will see it deployed more broadly by military institutions over the coming decades.

## DISRUPTOR FOUR: NATURAL THREATS
## AND THE ANTHROPOCENE

It would be difficult to examine the nature of change in the global security environment without some examination of climate change and other natural threats. Overwhelming evidence exists that the Earth's climate is changing, and that these changes—at least in many of the areas we can measure—are outside long-term norms for climate variation.

The 2004 book *Global Change and the Earth System* examined the many impacts human activities have on global climate and Earth systems. The book also featured a series of graphs that mapped human-driven change since the first Industrial Revolution. These graphs chart a series of human endeavors since 1750 that have demonstrated explosive growth that includes water use, fertilizer consumption, motor vehicles, and telecommunications, as well as areas such as carbon dioxide in the atmosphere and biodiversity. The graphs have subsequently been used in various publications including *New Scientist* and Thomas Friedman's book, *Thank You for Being Late*.[136]

Not long after the publication of *Global Change and the Earth System*, the phrase *great acceleration* gained usage in elements of the scientific community. Emerging from a conference in 2005 on the relationship between humans and the environment, the term has been applied by scientists to describe the degree of change in many socioeconomic and Earth systems trends.[137] This accumulating series of accelerations, beginning with industrialization in the 1750s, has led many scientists to advocate for the acceptance that we are now in a new geological epoch, which has been called the *Anthropocene*. The term has gained wider acceptance since Paul Crutzen and Eugene Stoermer proposed it in 2000 to describe the current geological era in which many processes on Earth have been altered by human impact. They proposed that "considering these and many other major and still growing impacts of human activities on earth and atmosphere, and at all, including global, scales, it seems to us more than appropriate to emphasize the central role of mankind in geology and ecology by proposing to use the term 'Anthropocene' for the current geological epoch."[138]

Despite the popularity of the term with scientists, the media, and scientifically engaged members of the public, it is not yet an official geological epoch. An important step in this direction was taken by the 2019 Anthropocene Working Group of the International Union of Geological Sciences, whose members voted overwhelmingly to propose the acceptance of the term at a

2021 meeting of the Executive Committee of the International Union of
Geological Sciences.[139]

————

In 1988 the United Nations Environment Programme and the World
Meteorological Organization established a new body charged with bringing together the best scientific research on climate change in order to provide policymakers with analysis and options for mitigation and adaptation.
This Intergovernmental Panel on Climate Change (IPCC) has more than
190 member nations. It delivers regular assessments written by hundreds
of leading scientists, examining key climate change markers that include
the oceans, heat (on land and in the atmosphere), drought, ice, and greenhouse gases.

Since 1990 the IPCC has also undertaken major assessments that review
scientific evidence for change, risks, and adaptation/mitigation options. The
assessment reports are delivered every five to six years, with the most recent
report—"AR5 Synthesis Report: Climate Change 2014"—coming in 2014.
Some of its key findings noted that "anthropogenic greenhouse gas emissions have increased since the pre-industrial era, driven largely by economic
and population growth, and are now higher than ever. This has led to atmo-
spheric concentrations of carbon dioxide, methane, and nitrous oxide that
are unprecedented in at least the last 800,000 years."[140]

But what does this mean for climate and for human civilization? The report
makes findings that inform governments and policymakers about the potential impacts of this shift in the climate. Among the current impacts are
the melting of glaciers, large-scale erosion, the reduction or destruction of
land and marine ecosystems, impacts on food production brought about
by floods and drought, and increased chances of wildfires. Over the course
of the twenty-first century, surface temperatures are likely to continue rising, resulting in more and longer heat waves and changes in precipitation in
many regions, which will impact food production and human health. This is
also likely to be accompanied by rising seawater temperatures.[141]

Notwithstanding this collaboration of thousands of scientists around the
world to provide evidence of changing temperatures and weather patterns,

the topic remains divisive and subject to much political discussion and derision. Despite the efforts of those who challenge the science or the politics of climate change, or who claim that adaptation and mitigation are too expensive, the evidence of a shift in global climate patterns is well established through the efforts of thousands of scientists worldwide over many decades. Individuals' rights to disagree and to challenge evidence must be respected. But in the case of climate change, to a lay person like myself, the evidence for action is as compelling as it is possible to be.

The recent COVID-19 pandemic represents a different kind of natural threat. While the pandemic presents no new forms of disruption—disease has been with us throughout human history—it has accelerated some of the change that is examined throughout this book. War and pandemics are two of the great scourges of humankind. For our entire history, humans have been threatened by each other and by the microscopic dangers that represent disease. As one commentator wrote at the height of the recent pandemic, "Biological shocks have been a persistent force of disruption in human history—destroying empires, overthrowing economies, decimating entire populations. Particularly when they spark or coincide with other crises—climate crises, legitimacy crises, monetary crises, and armed conflict—they mark moments of transformation or redirection in the stream of history."[142]

The 2020 COVID-19 pandemic has also had the effect of accelerating—for a short period—change in the global order and types of security challenges that nations must address. It is also likely, however, to speed up innovation in many areas including health care, online learning, and manufacturing supply chains. Nations have demonstrated a newfound desire for greater sovereign resilience. This will require long-term strategies complemented by the capacity to rapidly adapt to these foreseen but still surprising changes in the environment. Progress in this area will be relevant to future pandemics but also to dealing with other natural disasters and long-term climate change.[143]

## IMPACTS OF DISRUPTION

For military institutions, these disruptive trends will have a range of impacts, both specific and overlapping.

## *Geopolitics*

Strategic competition between large powers will dominate geopolitics for at least the first half of the twenty-first century. Competition is something different from war. As Tom Mahnken has written, "Competition lies midway on the spectrum whose ends are defined by conflict and cooperation."[144] A recent U.S. Marine Corps doctrinal publication called *Competing* explains competition as "states and non-state actors seek[ing] to protect and advance their own interests. . . . Competition results when the interests of one political group interact in some way with those of another group. These interactions take place in a dynamic environment. Each move an actor makes toward fulfilling an interest changes that ecosystem. Any interaction of interests changes the situation as well."[145] This new era of strategic competition is just the most recent case of large and powerful nations competing for influence and supremacy over their neighbors. However, it will drive the design and readiness of most military forces in the coming decades.

In studying the evolving strategic competition, one can learn from the previous strategic competition between superpowers. The Cold War between the United States and the Soviet Union featured peaks and troughs in tension. One of the worst peaks was the Cuban Missile Crisis, where Soviet commanders on at least one occasion came close to using tactical nuclear weapons against their American foes.[146] But the Cold War also featured areas of cooperation—the Apollo-*Salyut* program is just one example—as well as an evolving appreciation of the need to agree on limits for conventional and nuclear weapons.

There was also no shortage of smaller conflicts during the Cold War. The Korean War (1950–53), the Vietnam War, the wars between Israel, Egypt, and Syria, the Iran-Iraq war, and the Afghan war may not have involved direct confrontation between the United States and the Soviets, but they both contributed troops, supplies, and support in almost every case. And this is the lesson that we must heed: while a major war between the United States and China might be avoided in the coming decades, avoiding smaller-scale conflicts involving proxies will be more difficult.

As we have seen throughout history, small wars and conflicts beneath the threshold of major wars can involve a mix of different actors. State-based military organizations may be players in this environment, but nonstate actors—which include hackers, private military companies, and extremist organizations (both state-sponsored and otherwise)—will all feature in this environment. Contemporary examples include the recent conflict in Syria and Ukraine.

War and competition in the twenty-first century will place a premium on the institutional ability to adapt between different types of missions and to do so rapidly. It will demand the capacity to quickly evolve new concepts for the physical domains of war as well as the cyber and information fields. It will require military institutions to reconsider their ideas, their institutional constructs, and how they prepare their people for the challenges of twenty-first-century competition and conflict.

## *Demography*

Demographic shifts will have several impacts on military activities. First, aging populations will reduce the percentage of national populations available in the broader workforce and the availability of young people for military service. This is a challenge for recruiters and the maintenance of sufficiently sized military organizations. It is likely to drive greater automation, in turn resulting in changes in training and education to support a more integrated human-machine military institution.

Smaller populations combined with aging populations magnify the challenge of enough working-age people to fill jobs. Importantly, it complicates the recruiting of sufficient numbers of quality personnel into military organizations, and then retaining them once trained. It will require more innovative ways of attracting the right people and potentially more partnerships with industry to share those with the most critical and in-demand skill sets.

Higher urban populations mean that military operations are much more likely to take place in cities. This will require a shift in emphasis in military operating concepts, as well as the training and education of military personnel. Finally, migration is a double-edged sword. On one hand, it can aid aging nations through the influx of younger people, who can contribute to the national economy through work and can supplement the number of people available for military service. On the other hand, a large number of migrants can overwhelm governance and economic systems of some nations, potentially resulting in unrest or conflict.[147]

## Technology

The development and proliferation of disruptive technology will continue to shape military institutions and their activities, as it has historically. However, technology has now become a much more level playing field. The 2017 U.S. *National Security Strategy*, the 2021 U.S. *Interim National Security Strategic Guidance*, and the 2017 and 2018 *Defense of Japan* government publications have acknowledged the potential for their highly sophisticated military institutions to face technological parity against future adversaries.[148]

This leveling trend is likely to continue as advanced technologies such as robotics, artificial intelligence, biotechnologies, information warfare, cyber, and space capabilities proliferate. New and disruptive technologies are becoming more accessible to nonstate actors and private military companies. The decline, or even loss, of this edge has considerable implications for how military institutions might consider future operations and intellectually prepare military personnel to lead and plan in this rapidly evolving environment.

Colin Gray has written that "every new set of technological marvels brings with [it] specific novel challenges. For every shiny new solution, we discover new problems. The principal reason this is always so is because of the inconvenience represented by the enemy."[149] This also means that technological developments in different military organizations will exist within a competitive environment and reinforces the likely transience of future technological advantages.

## Climate and Natural Threats

Changes in climate and other natural threats will impact the kinds of missions that military institutions will undertake in the coming decades, while being the principal design driver. The most recent assessment of future trends by the United Kingdom's Defence Concept Development Centre has found that climatic shifts will have far-reaching consequences over the next thirty years, with floods, droughts, storms, heat waves, and heavy rainfall all expected to become more intense and possibly more frequent. Climate change may also exacerbate migration and security challenges. For example, long periods of low rainfall have been a driver for increases in violent conflict, because of scarce vital resources.[150]

Global environmental and ecological degradation, as well as climate change, is likely to fuel competition for resources and cause economic distress and social discontent through the 2020s and beyond.[151] It will remain a concern for future military leaders, strategists, and national policymakers to ensure that they are able to respond to the worst effects of climate change. Whether it is widespread natural disasters, the migration of refugees from affected areas, or even the competition for resources in a world changed by different climate patterns, military institutions must be prepared for the kinds of missions that they will confront in this changed environment.

––––––

These trends are evolving quickly. At times, keeping pace with the amount of change is very challenging for institutions, let alone individual humans. The interaction of the various disruptive elements also complicates developing solutions. One solution that nations may reach for, which nations since antiquity have embraced, is conflict rather than cooperation. This is not to say it is a preferred solution. But it will continue to feature as a potential avenue for nations and their leaders to pursue in balancing their response to these global trends and the demands of their people. We now turn to examining how these trends impact warfare and the future of human competition.

# 2

## FUTURE WAR

### *New-Era Competition and Conflict*

> Warfare has re-invaded human society in a more complex, more extensive, more concealed, and more subtle manner. While we are seeing a relative reduction in military violence, at the same time we are seeing an increase in political, economic, and technological violence. The new principles of war are no longer using armed force to compel the enemy to submit to one's will, but rather are using all means, including armed force or non-armed force, military and non-military, and lethal and non-lethal means to compel the enemy to accept one's interests.
>
> —*Qiao Liang and Wang Xiangsui*[1]

In the early 1930s a Japanese naval officer set out to explore the challenges that he believed would shape future warfare. Tota Ishimaru sought to write a book that would complement works on future conflict by European authors. But he also intended to challenge what he saw as an idea fixed in the minds of Western writers: that "Japan and Germany are disturbers of the peace."[2]

Ishimaru's book explored earlier wars and their sources, as well the foundations of future conflict in both Europe and the Pacific. *The Next World War* provided useful insights from a military professional on the state of international relations and trends driving international tension such as economic blocs and anti–status quo forces. It also examined how these might manifest as warfare in Europe and the Pacific.

In retrospect, the book was prophetic. Ishimaru described how Japan in the mid-1930s possessed a strong resemblance to Germany before World War I. He wrote that "the Germans were intoxicated with success; they became so conceited as to imagine themselves the most superior people in the world. They regarded war as a means of national aggrandizement, bringing eventually their country to the test of the Great War."[3] He concluded his book with a warning: "If the Japanese people should become overconfident because of their modest successes, they may wake up one day to find themselves facing the whole world as their enemy; and they will then meet the same fate as did Germany."[4]

Prophecies that attempt to anticipate the location and type of future conflicts are hardly new. The Greeks and the Romans both linked war with the intrigues of the gods.[5] They accepted that war was an element of human existence as they understood it and that it would inevitably be part of their future. It remains the responsibility of governments and senior military leaders to anticipate future conflicts and to build their nations' military power to deter or respond to threats.

The aim of this chapter is to examine how the new technologies and the changed geopolitical, demographic, and climatic environment reviewed in the previous chapter will transform war. This knowledge should inform contemporary and future military leaders at all levels so they can build and apply military power in the achievement of political outcomes. It also provides a foundation for the following chapters that examine the future of military ideas, institutions, people, and the holistic application of military power.

In every age there are those who try to make the case that humans today are more enlightened than their predecessors and less likely to stumble into war. Whether it is Norman Angell before World War I or more modern prophets such as Steven Pinker, there will always be those who propose that human beings no longer need to make war on each other.[6] I really want them to be right. But the truth is that our future is more likely to resemble the past. As Colin Gray has written, "It is a reasonable assumption that future strategic history will resemble the past and present. Because it rests upon the evidence of 2,500 years, this is not a recklessly bold claim."[7]

Therefore, we must invest in understanding war—how it evolves and how it remains the same. Those who study war and the military profession know this as war's *changing character* and *enduring nature*. This notion of how war both changes and stays the same originated, at least in written form, with Carl von Clausewitz's *On War*. It is still a key framework for the contemporary study of war as a phenomenon.

The enduring nature of war refers to those aspects of human conflict that are consistent themes throughout the ages. Clausewitz produced his masterpiece just as the first Industrial Revolution started to transform warfare. His theories on the political objectives of war, the concept of friction in war, and the relationship between the government, the people, and the military (known as the *remarkable trinity*) possess an enduring relevance to military professionals. But this is also important for other national security professionals and national government leaders to understand. That wars are ultimately political, and that they are full of chance, uncertainty, and friction, is their enduring nature.

Military professionals and academics, informed by their study of Clausewitz and other theorists, recognize a difference between the objective nature and subjective character of war. Of significance is that war's nature captures those elements that differentiate it from all other human activities as a distinct phenomenon. This *nature of war* is defined by the interaction between opposing wills, violence, and political motivation.[8] In describing this theory of the nature of war, Clausewitz sought to depict what he believed were the universal—or enduring—elements that every theorist should be concerned about above all.[9] Doing so would "give the mind insight into the great mass of phenomena, then leave it free to rise into the higher realms of action."[10]

The other side of this coin is the changing character of war: a recognition that technological, political, and societal change drives innovation and adjustments in how war is waged. Humans have moved from sticks and rocks, to swords and shields, to horse-borne warfare, and now into the age of machines that are used across all domains of competition and conflict. The ideas of war have also developed in parallel with the tools of war throughout the ages. This process continues today and is an important component of war's changing character.

The current pulse of global change is seeing humans' power to use machines being supercharged by the power of algorithms and data. It has a variety of impacts on war, which will then flow into how military institutions, and their nations, think about and prepare for war. But the reasons why humans fight have remained relatively stable over thousands of years. Athenian general and historian Thucydides wrote of how "fear, interests, and honor" drove Athenian strategic decision-making.[11] While we may use different language in the contemporary world, those impulses represent continuity in warfare. We will examine this continuity in warfare first to provide a baseline for a subsequent investigation into how war is changing.

## WAR

The old Roman proverb "If you want peace, prepare for war" is representative of how warfare has not changed. In his 1983 book *The Causes of War*, Sir Michael Howard described how "it is hard to deny that war is inherent in the very structure of state."[12] It is an element of continuity in the matters of states and in human affairs. There is a certain timeless quality about humans' pursuit of "fear, interests, and honor" through warfare. Regardless of the era, throughout recorded history, people have competed and fought over what they desired, as either individuals or societies. It is therefore important to appreciate the continuities in warfare as a foundation for our exploration of future forms of warfare.

We need to appreciate these continuities for several other reasons. First, while technological change is advancing rapidly and geopolitics constantly evolves, some aspects of society and the wider security environment either do not change or change at an almost imperceptible pace. An example of this might be human nature.

Second, appreciating continuity provides part of the context against which future military and national strategies are developed, refined, implemented, and adapted. The strategic cultures of China, Russia, and the United States display a range of constants over time. An example of this might be the 2019 U.S. strategy for the Indo-Pacific. The start points for the strategy are long-standing aspects of the U.S. military approach to defending the western Pacific, as well as its relationships in the region.[13] Against these more enduring elements are cast the key aspects of change in the direction of U.S. strategy in this region.

Third, understanding those elements that are not likely to change assists in focusing the doctrine, training, and educational programs of military organizations. These institutions, and other elements of the national security community, might then apply this knowledge to balance their investment in the intellectual development required to address change in the environment with that invested in enduring skills and knowledge requirements.

Fourth, understanding the enduring features of war provides a framework for detecting changes in the security environment that might impact how we conceive of war. Detecting change is an important aspect of strategic leadership and of strategy development and implementation for military institutions. Being able to recognize change in the modalities of war allows institutions to adapt their ideas, equipment, and organizations as well as their training and education.

Finally, there is bias in many contemporary books and reports toward examining change. While important, it often means that unchanging elements are either overlooked or downplayed in importance. As Stephen Biddle has written in *Military Power*, "Change of course is inevitable. But so is continuity. And today's policy debate systemically exaggerates the former and slights the latter."[14] Therefore, a thorough appreciation of war's continuities offers an important foundation for studying the future of human conflict and competition.

The first and most important continuity is that war will remain a part of human affairs well into the future. In his study of war across several millennia, *War in Human Civilization*, Azar Gat notes that "the solution to the enigma of war is that no enigma exists. Violent competition is the rule throughout nature. Humans are no exception to this general pattern."[15] For thousands of years of recorded history, humans have sought to impose their will on each other. Over this time, humans have expressed their creativity and brutality through the conduct of warfare. There is little evidence that the presence of fear, honor, and interests in human decision-making has changed.

More recently, in writing about the impact of warfare on the development of human societies, Ian Morris has described war as "something that cannot be wished out of existence, but that is because it cannot be done." He further describes war as the only method humans have found for explaining the decrease in violent death rates of 10 to 20 percent in small prehistoric groups to below 1 percent in today's globalized society. War has made the planet peaceful and prosperous—so prosperous that "war has almost but not quite put itself out of business."[16]

Morris's proposition also reflects the modern manifestations of the decline of war theory, which has always enjoyed some popularity among academics and strategists. Academics such as Bethany Lacina, Nils Petter Gleditsch, and John Mueller (among others) have written on the decline (or obsolescence) of war.[17] But in recent years, books such as Steven Pinker's *The Better Angels of Our Nature* have become bestsellers, and this hypothesis about war's decline has regained a certain level of popularity.[18]

The core argument of the decline of war theorists such as Pinker goes like this: human nature comprises inclinations both toward and against violence. Over time, the counteraction against violence has gotten stronger, and there has been a corresponding reduction of violence, including war, insurrections, murder, domestic violence, and child abuse. Across six chapters, Pinker explores the trends that are related to declining violence:

1. The Pacification Process: This is the transition from "the anarchy of hunting, gathering, and horticultural societies to the first agricultural civilizations with cities and governments," resulting in a decline in raiding and feuding and a "fivefold decrease in rates of violent death."[19]

2. The Civilizing Process: Pinker argues that "between the late Middle Ages and the twentieth century, European countries saw a tenfold-to-fifty-fold decline in their rates of homicide."[20]

3. The Humanitarian Revolution: Pinker writes how this revolution "unfolded on the scale of centuries and took off around the time of the Age of Reason and the European Enlightenment in the seventeenth

and eighteenth centuries." This revolution resulted in the first organized attempts to abolish slavery, judicial torture, sadistic punishment,
and cruelty to animals.[21]

4. The Long Peace: This took place after World War II, where the major
   powers of the world and many developed states stopped waging war
   against each other.[22]

5. The New Peace: Even Pinker recognizes this trend as tenuous, but its
   essence is that since the end of the Cold War, organized conflicts, including civil wars and genocide, have declined throughout the world.[23]

6. The Rights Revolutions: Pinker describes a post–World War II period
   that has resulted in "a growing revulsion against aggression on smaller
   scales, including violence against ethnic minorities, women, children,
   homosexuals, and animals. These spin-offs emerged from the concept
   of human rights—civil rights, women's rights, children's rights, gay
   rights, and animal rights."[24]

Declining violence and a corresponding decline in deaths from battle and
other forms of violence in the world have many benefits—for individuals and
for societies. No sane person could wish otherwise. But is this likely to be the
case? There are challenges to the decline of war theory from multiple vectors.

Anthropologists have challenged Pinker's findings, disputing the assumptions about the level of violence in prehistoric societies. Anthropologist
R. Brian Ferguson rejects Pinker's findings, asserting that his book contains a "selective compilation of highly unusual cases, grossly distorting war's
antiquity and lethality."[25] Other critiques have been related to postmodern
approaches or the types of data used by Pinker.[26]

In a 2016 paper, Pasquale Cirillo and Nassim Taleb also critiqued Pinker's
thesis about the decline of war. Reviewing data from a range of sources on
human conflict since AD 1500, they determined that no particular trend in
the number of armed conflicts can be deduced. They conclude that "at least
in the last 500 years humanity has shown to be as violent as usual. . . . One
may perhaps produce a convincing theory about better, more peaceful days
ahead, but this cannot be stated based on statistical analysis—this is not
what the data allows us to say. Not very good news, we have to admit."[27]

Another rebuttal to Pinker's work worth reviewing is from Bear Braumoeller in his 2019 book *Only the Dead*. Braumoeller's most effective criticism focuses on the "data and sound reasoning" applied by Pinker. Examining different statistical techniques used by political scientists in the study of international conflict, Braumoeller explains the difference between significant and anomalous findings. His analysis rebuts Pinker's thesis and finds that any decline in the deadliness of war is within the normal range of variation, with no downward trend over the past two hundred years.[28]

The reality is that with several thousand years of recorded history, warfare has been a continuous aspect of human interactions. While war might be less frequent in the future, national and military leaders would be irresponsible to use this decline of war theory to justify running down their armed forces; in doing so, they might well be setting the conditions for another war instead of deterring it. Deterrence is still a compelling theory in international security affairs. As Braumoeller writes, "Humanity seems most inclined to work for peace when the danger of war is most apparent—typically in the aftermath of huge wars. Fostering the belief that war poses no threat is a good way to convince people not to prepare for it. If warfare is indeed on the decline, we can afford to react minimally to each of these potential threats in the hopes of avoiding a lethal spiral of escalating conflict. If it is not in decline, however, our passivity might cost us dearly."[29]

A second continuity is that human competition is enduring, with war being the violent manifestation of this ongoing struggle. Human competition is never going away; it may recede at times (think the post–Cold War era) but is a constant feature of the interplay of nations, groups of nations, and nonstate entities.

The new era of strategic competition was explored in the previous chapter. Viewing this interaction of nations as a competitive environment allows the development of strategies for success or at least for achieving a minimal set of national objectives. But for nations to appreciate this competitive environment and work within it successfully, they require several important elements of knowledge.

First, they must understand themselves, including their national strengths, weaknesses, biases, and potential sources of advantage. Second, they must understand the other actors within this competitive environment, including their strategic culture, objectives, and sources of strength and potential weaknesses. These other actors are not always nation-states; they might also be major corporations that possess strategic influence (think Google or Facebook), multinational organizations such as the United Nations, or non-state actors such as ISIS. Third, nations must understand the various dimensions of competition—where the areas of interaction are, where they might productively interact, and how they might learn from these interactions. They might occur in the various domains of diplomacy, information, technology, and culture. Finally, nations in this competitive environment must possess some theory of victory—a notion of what they want and the accompanying strategies to achieve it.

There has been a profusion of models that seek to provide a framework for the current era of reemerging great power competition. Descriptions such as "cooperation, competition, and conflict," "coercion, collaboration, and conflict," political warfare, and gap war are among the contemporary attempts to intellectually come to grips with how national security affairs are rapidly evolving.[30] There is nothing wrong with any of these descriptions, and each is useful for specific institutions and parts of the challenge.

It would considerably simplify things if we were to view the world as it is: an environment where competition is constant. This is the historical norm. Competition takes many forms—economic, cultural, informational, diplomatic, and violent. However, when we coin new buzzwords such as grey zone or effects-based operations, we also introduce poorly defined terminology that obscures rather than clarifies meaning.

In a 2020 article examining this phenomenon of obscuration through arcane lexicon, Donald Stoker and Craig Whiteside note that this buzzword-laden sophistry is dangerous. They write that such new terms "should be eliminated from the strategic lexicon. They cause more harm than good and contribute to an increasingly dangerous distortion of the concepts of war, peace, and geopolitical competition, with a resultant negative impact on the crafting of security strategy for the United States and its allies and partners around the world."[31]

Many members of the profession of arms, as well as those in academic think tanks, have been drawn to buzzwords and fad concepts that promise quick or easy victory through new technologies or new concepts. We must return to a more accessible theoretical construct regarding competition. This is not to say it is simple. Clausewitz wrote that "knowledge in war is very simple. But that does not make its application easy."[32] Nonetheless, we have to do this. In making our theoretical construct of competition more easily accessible by a larger proportion of our people, we not only provide greater transparency; we also allow for the examination, critique, and competitive tension that can only improve our strategies for success within this strategic environment.

Plain English should be used more in military institutions to describe the competitive environment in which we exist. As a member of the military profession, I am the first to admit that we have a long way to go. We must improve how we communicate with each other, our governments, and our people. Not only does clear communication allow for better conversations across the civil-military divide—what Eliot Cohen has called the "unequal dialogue"[33]—it also builds trust in military institutions by government and the people because they can understand how military organizations describe themselves and their institutions. This is vital in democratic societies.

We should clearly describe and communicate the form of competition we are currently in. While human competition is a continuous feature of existence on this planet, the form of competition in the twenty-first century is somewhat different from previous eras. This differentiation is driven by new means to disseminate information (the Internet, social media, big data analysis, and sophisticated algorithms) and by new, long-range nonnuclear weapons systems that provide a layer of conventional deterrence within the overall competitive environment.

A third continuity is that military institutions will still exist. Beatrice Heuser notes that "something that has not changed is the close attachment of armed forces to states, making them a particularly easy instrument for states to control and use."[34] Governments, even those without strong inclinations toward lavish investment in the military, appreciate how military institutions can respond to and achieve outcomes across a range of different situations.

However, military institutions, particularly those that possess a network of external relationships from which they can learn, are unlikely to remain static organizations. Military institutions will continue to adapt as technology, society, and geopolitics evolve. As Trevor Dupuy notes in his 1984 study of the progress of weapons and warfare through the ages, the story of warfare is an account of continual change.[35] The necessity to learn and adapt has been the subject of multiple accounts that have examined military failure, military adaptation, and revolutions in military affairs. However, not all military institutions have been successful at learning. They are inherently conservative organizations, with long-term membership and strong assimilation mechanisms.[36]

Military organizations have deep institutional memories about their successes and failures. Those memories do not necessarily always align with historical facts. But they do drive their individual cultures and, coupled with a national defense strategy, determine the institutions' outlooks. Additionally, different services and military occupational specialties develop unique subcultures within the larger force. These cultures and subcultures are powerful elements in developing cohesion and esprit de corps, but they can also be barriers to change. The tighter and more cohesive these subcultures become, the more entrenched their ideas and processes become, and the more they resist the adoption of new ideas, techniques, and technologies.[37]

There are many case studies of how institutional culture has proved an obstacle to necessary change in military institutions. One fine example that illustrates this is the transition of the U.S. Army from horses to tanks in the late 1930s and early 1940s. For a country that invented the first modern production line for motor vehicles and was the leading economy of the early twentieth century, it is hard to believe that its army would hold onto its horses well into the age of mechanization. But hold on it did, as former military officer David Johnson explains:

In the U.S. Army, mechanization became captive of the conservative infantry and cavalry branches, which saw the technology through the lens of improving what they already did. Or worse, as in the case of Maj. Gen. John Herr, the last chief of cavalry, who actively blocked mechanization to maintain horse cavalry structure. His approach to mechanization is best summed up in his statement: "When better roller skates are made, Cavalry horses will wear them." His breakthrough innovation was to put horses on tractor-trailers to give them operational mobility. Herr ran the cavalry branch until Army Chief of Staff Gen. George Marshall got rid of all the Army's horses—and Herr—in 1942.[38]

There are, however, many examples of successful innovation and change in military institutions. Academics and historians such as Williamson Murray and Allan Millett,[39] Dima Adamsky, Colin Gray, Stephen Rosen, Frank Hoffman, Dave Barno, Nora Bensahel, and Michael O'Hanlon have offered historical insights into how military institutions have adopted new ideas and technologies in the past and how they might institutionalize such a process.

The adaptive capacity of military institutions is also influenced by the adaptive capacity and cultures of adversaries and competitors. As Gen. Robert H. Scales Jr., USA (Ret.), notes, every successful innovation that provides military advantage eventually yields to a countervailing response that shifts advantage to the opposition.[40] Future national leaders and military leaders will need to appreciate the competitive and adaptive nature of warfare. They must be able to recognize the need for change and lead that change where necessary. Frank Hoffman, in his 2021 book *Mars Adapting*, describes how "military forces have always had to learn how to learn and then use that knowledge to change in order to succeed."[41] And as Nora Bensahel and David Barno describe in their 2020 book *Adaptation under Fire*, "Adaptability is one of the most, if not the most, important attributes of military forces."[42] We will explore this concept of an adaptive approach more in chapter 3.

A fourth continuity is that nations will continue to seek to protect their sovereignty and, in doing so, force will remain subordinate to political requirements. While we might reframe the nature of national sovereignty in a world connected through cyberspace and commerce, protecting national sovereignty remains an enduring responsibility of governments. An important aspect of this is national security. Nations will approach this in different ways depending on their individual strategic cultures.

The strategic culture of a nation is a significant element of understanding continuity in how it protects its interests. It is the context that surrounds and supplies meaning to strategic behavior, which is adopted by culturally shaped or *encultured* people, organizations, procedures, and weapons.[43]

Colin Gray, who examined the concept of strategic culture at length, argued that understanding strategic culture could provide an improved capacity for understanding enduring policy motivations and make predictions, as well as understand the meaning of events in the assessment of others.[44]

For most countries, protecting their sovereignty and their interests will still involve possession of the various elements of national security, including intelligence services, foreign affairs departments, alliances and treaties, standing military organizations, and mechanisms to blend the outcomes of these entities in the achievement of national objectives.

Beyond military forces, one of the most important mechanisms states use to achieve national objectives is strategy. States will continue investing in the development of national and military strategies. In protecting their sovereignty and seeking national goals, strategy-making by nation-states will remain an enduring element of governing and national affairs. As Colin Gray writes, it is "the glue that holds together the purposeful activities of state."[45]

Over the past several decades, strategy-making has become a more integrated undertaking, described in the United Kingdom as a "joined-up" approach. MacGregor Knox has noted that strategy-making remains "the domain of states, those remorseless monsters whose central characteristic is the monopoly of violence on their own territory and whose pivotal institutions are armed forces and bureaucracy."[46] Further, he depicts bureaucracy as a central element in the maintenance of armed forces and the development of national policies and strategies. However, bureaucratic systems have not always been powerful enough to impose strategic coherence. According to Knox, "The most elaborate military bureaucracies in history retain a wider preserve of absolutism at their summit and undiminished scope for bureaucratic autism below."[47]

There are many strategies a nation might adopt. Some, particularly small states, will adopt defensive postures, focus their attention on their immediate region, and possibly nurture alliances with larger and more powerful states. Other states have adopted defensive strategies, keeping a regional focus but choosing not to formally ally themselves with another nation, taking a neutral stance (think India and Switzerland). The largest and wealthiest of states have embraced more balanced strategic postures, with a global outlook and a vast network of alliances and security partnerships. All states, regardless of their size, will use a large proportion of their national resources, including diplomacy, information, culture, military, and economic power, to deter others or compel them to act in a way that allows the achievement of their strategic goals.

Because of the different sizes and interests of nations and their differing capacities to resource their national security needs, war in the twenty-first century will also remain a mix of different techniques and approaches. During the Cold War, the key players, including the United States, the Soviet Union, and European countries, invested predominantly in conventional forces, information activities, and strategic capabilities. This did not mean, however, that they and other state and nonstate entities did not engage in low-level violence.

During this global strategic competition, smaller wars and other hostile activities between the United States and the Soviet Union (and their proxies) took place. These included the Korean War, the Cuban Missile Crisis, the Vietnam War, the Soviet invasion and occupation of Afghanistan (1979–89), and the shooting down of each other's aircraft.[48] This too will be the case in the new strategic competition between the United States and China.

A fifth and final continuity in war (and competition) is surprise. This is usually defined as striking at a place or time in a manner for which an adversary is not prepared.[49] Lawrence Freedman has written that "a surprise attack, conceived with cunning, prepared with duplicity and executed with ruthlessness, provides international history with its most melodramatic moments. A state believes itself to be at peace then suddenly finds itself at war, in agony and embarrassed that it failed to pick up the enemy plot."[50]

This approach, embraced in both the Eastern and Western traditions of war, has aimed to surprise an adversary and overwhelm them when they are at their weakest or when they least expect it. Sun Tzu wrote that forces should "go forth where they do not expect it, where they are not prepared."[51] In Russia, the word *maskirovka* describes the range of military activities designed to deceive, deny, and camouflage friendly intentions. While achieving its pinnacle in operations against the German army in 1944–45, this philosophy continues to underpin contemporary Russian operations in the Crimea and Ukraine to achieve strategic political outcomes as well as tactical objectives on the battlefield.[52]

Humans have been endlessly surprised, over thousands of years, at all levels
of war. One of the earliest recorded examples is from the Egyptians in 1294
BCE. Pharaoh Ramses II of Egypt was surprised by his enemy when he was
led into an ambush by two Hittite "deserters" while leading his army against
the Hittite stronghold of Kadesh.[53] Throughout the writings of ancient historians such as Thucydides, Polybius, and Herodotus are multiple examples
of one combatant gaining victory through surprise and deception.

British military officer, writer, and military theorist J. F. C. Fuller wrote
extensively on surprise as one of his principles of war. In his classic 1926
book, *The Foundations of the Science of War*, he notes that "in war, surprise
is omnipresent; wherever man is there lurks the possibility of surprise. If he
wishes to understand war, [he] must examine the nature of surprise in its
thousand and one forms. Surprise should be regarded as the soul of every
operation."[54] He further refined his thesis on the application of surprise by
classifying it into three categories: mental, moral, and physical.

In the twentieth century, attempts to achieve surprise have featured in the
operations of the main protagonists in all significant conflicts. Pearl Harbor
in 1941, the October 1950 Chinese entry into the Korean War, the Arab
attacks on Israel in 1973, and the opening attacks of Operation Desert Storm
in 1991 are a small sample of the successful achievement of surprise in war.

Military theorists have also distinguished between the levels—strategic,
operational, or tactical—at which surprise might be achieved. The level at
which the principal action is achieved, however, may be different. For example, on D-Day, strategic and tactical action relied upon operational surprise. More recently, on 9/11, strategic surprise was generated by the tactical
actions of nineteen terrorists and the four civilian airliners they hijacked.

In *Pearl Harbor: Warning and Decision*, Roberta Wohlstetter offers a comprehensive analysis of surprise in warfare through an examination of the
attack on Pearl Harbor:

Surprise, when it happens to a government, is likely to be a complicated, dif-
fuse, bureaucratic thing. It includes neglect of responsibility, but also responsibility so poorly defined that actions get lost. It includes gaps in intelligence,
but also intelligence that like a string of pearls too precious to wear, is too

sensitive to give to those who need it. It includes the alarm that fails to work, but also the alarm that has gone off so often it has been disconnected. . . . It includes the inability of individual human beings to rise to the occasion until they are sure it is the occasion—which is usually too late.[55]

Contemporary military institutions also seek to achieve surprise against their adversaries. The doctrinal publications of military organizations in the United States, the United Kingdom, Australia, and other countries describe achieving surprise against adversaries as a highly desirable element of military strategy and operations.[56] However, the methods of creating surprise have expanded over the last two decades. New areas such as high-speed hypersonic weapons and cyber operations now supplement traditional means of deceiving an adversary and achieving surprise. But the breadth of endeavors in a national security enterprise has also broadened. Therefore, surprise might be generated outside the military domain as well.

The desire to achieve surprise against an adversary, to place them in a position where they are weaker than one's own forces, has been an enduring feature of warfare. It is almost certain to be a core area of study for future military leaders to allow them to minimize the chances of their own forces being surprised and to enable them to rapidly respond and adapt if they are.

## THE CHANGING CHARACTER OF
## WAR IN THE TWENTY-FIRST CENTURY

In February 2014 Russia annexed the Crimea from Ukraine. The Russians then executed a campaign to weaken Ukraine by supplying and organizing separatist groups in the east of the country while waging a wider strategic influence campaign in Europe and beyond. Russia had originally sought to cloak its involvement by using proxy forces. But as the conflict escalated in 2014 and 2015, deniability on the part of the Russians became increasingly difficult. Over succeeding years, Russian proxy forces, Ukrainian forces, and Ukrainian proxies have fought a bitter struggle in and over Ukraine, in cyberspace, and in the minds of populations in Ukraine, Russia, and beyond.

The war in Ukraine comprised a combination of old and new techniques. It has featured ancient elements of war such as the siege of cities and other key points (like the Donetsk airport), the negotiation (and breaking) of truces between the various combatants, the employment of state and nonstate combatants in physical combat and information warfare, and the conduct of psychological operations to erode the will of combatants and civilians.

The war has also featured more recent forms of warfare including the widespread use of unmanned systems married with electronic warfare and artillery/rocket systems to find, fix, and neutralize targets across the country. It has resulted in evolved organizations and conceptual approaches to the conduct of war, including the Russian battalion tactical groups, integration of information warfare at all levels, and more densely concentrated air defense artillery. The components of these approaches are often not new, but technologies and ideas have been combined and applied in new ways, particularly by the Russians and their proxy forces.

The war between Ukraine and Russia has also seen the involvement of third-party, open-source intelligence capabilities such as data mining, geolocation, and other methods. Bellingcat and Forensic Architecture are two examples of private companies involved.[57] This approach has evolved to incorporate the independent surveillance of conflict zones and detection of nondeclared state-based combatants and weapons systems. The pervasiveness of these actors in the contemporary environment highlights how transparent military operations have become to civilian surveillance and collection assets. It reinforces the need to understand, manage, and reduce the signature of military activities—from the tactical to the strategic levels.

Another form of third-party participant in Ukraine has been online hackers. Independent hackers and teams supported operations on both sides of the conflict. For example, in December 2014 Ukrainian hackers were able to access and then publish a cache of documents allegedly stolen from a Russian Ministry of Internal Affairs server. The hackers sourced the documents from various government departments and described Russian military casualties and other aspects of Russian operations in Ukraine.[58]

The Russian operations in Ukraine, which continue in 2021, and the supporting global information operations strategy build on decades of Russian developments in the physical, moral, and informational aspects of war. As

Dima Adamsky has written, the Russians have lacked a fascination with technology but early in the Soviet era created comprehensive guidelines and methods on how to discuss war scientifically.[59]

The more recent impetus for change began in 2008, when Russia commenced a series of military reforms that have aligned the size and capabilities of the military with the strategic environment, budget realities, and new technologies. The Russian *new-generation war* concept emerged through a process of trial and adaptation by the Russian state and its military forces. Often described by Western experts as hybrid warfare or cross-domain coercion,[60] its provenance stems from centuries of Russian warfare with its neighbors. In the past decade, the Russians have experimented with new military approaches in the Crimea, Ukraine, and Syria. More importantly, they have vigorously pursued new-era thinking.

The deepening reliance of the Russian Federation on cyber operations and influence activities is an important element of this evolved Russian thinking. It is apparent not only in their doctrine and operations in Ukraine and Syria but also in their attempts to subvert the U.S. elections in 2016.[61] The "active measures" of the Russian state to influence adversaries on the battlefield and in their homelands are likely to become more persistent, more sophisticated, and more difficult to detect and respond to.

The Russian new-generation warfare has several important pillars: the employment of nonmilitary instruments of national power in confrontation; broad use of asymmetric and indirect methods; systemic deception at all levels; winning conflicts cheaply without significant commitment to more expensive forms of warfare; and the pursuit of competitive advantages against much stronger adversaries. The center of gravity in these activities is much more nuanced than in traditional force-on-force competition; national and subnational populations become a key prize in an influence competition.[62]

In examining the Russian approach, Dima Adamsky has noted that in the ideal campaign, the "informational-psychological struggle" first takes a leading role, as the moral-psychological-cognitive-informational suppression of the adversary's decision-makers and operators assures conditions for achieving victory. Many aspects of these ideas have been included in the 2014 Russian military doctrine and operationalized in their military theaters such as Syria and Ukraine. General Valery Gerasimov has led this process. Along with many other Russian military strategists in the past decade, he has emphasized the broader application of nonmilitary measures across the spectrum of competition and conflict.[63]

Gerasimov has also emphasized that innovative thinking is fundamental to Russian strategy. He recognizes that creative ways of thinking will not emerge solely from policy-makers in military headquarters and the policy shops of government departments. He notes that "the involvement of all of the Ministry of Defense's scholarly organizations and the scholarly capabilities of interested federal organs of executive authority is required for a more effective development of these issues. It is necessary to discuss problem issues at scholarly conferences and examine them during 'round tables.' Only by doing this will new results in the field of the theory and practice of military strategy be achieved."[64] In Gerasimov's mind, strategy development is a team sport, and one that must be broadened to involve all the intellectual capacity of a state.

Russia has adapted to the fourth Industrial Revolution by building a new approach to securing its national security goals on the foundation of its traditional military culture. It is a form of warfare that seeks to manipulate an adversary's perception, to maneuver its decision-making process, and to influence its strategic behavior while minimizing (compared to the industrial warfare era) the scale of kinetic force use and increasing the nonmilitary measures of strategic influence. They have developed a skillful yet constantly evolving integration of informational, nonnuclear, and nuclear compellence and deterrence to achieve this.[65] Gerasimov has been reported describing this as a "strategy of limited action."[66]

Nations such as Russia and China have developed a sophisticated understanding of using military forces in different guises without public acknowledgment of hostile intent.[67] It is an important aspect of the changing character of war. David Kilcullen has described this in terms of *addition* and *combination*: "bringing into play the maximum range of categories of conflict and combining them in novel ways to pose integrated challenges that

an adversary may neither understand nor have the capacity to counter."[68]
National objectives that might historically have been decided by war are
increasingly decided in peacetime.

————

The Russian operations in Ukraine demonstrate that war remains an attractive option for political leaders to achieve their strategic objectives. But it
is a different form of war, and the Russian conceptual development and
operational testing of new ideas and tactics provide important insights for
future conflicts. In many respects, the conflict in Ukraine also shows how
war never really disposes of any old ideas or capabilities. It just combines
them in different and sometimes new ways depending on the aims and will
of the combatants. The first three industrial revolutions supplied a foundation of technologies, organizations, and ideas upon which our new era of
war is being constructed. The ideas and technologies from earlier eras intermingle with those from the present. Old technologies and organizations are
either reused or altered for different purposes.

Every innovation in the technology and ideas of warfare has built on the
foundation of preceding innovations. With the coming of steam engines,
horses were not all destroyed; their role changed. With the coming of the
combustion engines, steam-based propulsion did not go extinct. And with
the coming of computers, space flight, and the Internet, other forms of transport and communications were not arbitrarily extinguished.

This is also true for old ideas that do not relate to technology, such as influence operations. Seeking to influence the will of an army, the people, and the
political leadership of an enemy can be traced back to the first millennium
BCE. However, the means to do so have been continually updated through
the twentieth and twenty-first centuries. Whether used in tactical deception (honed to a high art by the Soviets and the Allies during World War II)
or more strategic information operations (used during the Cold War and in
the current era by China and Russia), new ideas about war and competition
are built on the aggregation of layers of old ideas, tactics, and organizations.

War as a phenomenon continues to evolve. New forms emerge through
experimentation, interwar innovation, and early war failures. New ideas,
institutions, tactics, and strategies that do appear provide a new layer on

the deep strata of existing knowledge about war. These existing strata compose the continuities we examined earlier in this chapter. To paraphrase Clausewitz, it is the "theory that exists so that one need not start afresh each time sorting out the material and plowing through it."[69] But this is just the basis for growth and change. As such, it is time now to explore how war will continue to change in the coming decades.

## TRENDS IN TWENTY-FIRST-CENTURY WAR

In the previous chapter, we explored how people who live through periodic surges in innovation possess a sense of accelerating change. The same is true with military organizations. Everything in war, and in preparing for war, seems to be speeding up.

New and disruptive technologies are being absorbed into society and military institutions. This changes how military institutions might conduct operations and undertake other military activities (such as disaster relief and humanitarian assistance at home and abroad) in the twenty-first century. At the same time, there are multiple continuities that will impact military operations. Building on this knowledge, we might now project how developments in the current industrial revolution may force change in the character of war. In the decades ahead, the most profound changes are likely to take place in seven aspects of conflict. While each trend might be examined as a disruption in its own right, the trends also interact and affect each other. These seven trends are as follows:

1. A new appreciation of time. The speed of planning, decision-making, and action is increasing due to hypersonic weapons, faster media cycles impacting political decisions, and the potential for AI to speed up decision-making at many levels. Concurrently, we are facing a long-term period of "cold but connected" competition and conflict with China. The core idea for future military organizations is to ensure that their people and institutions at every level are able to intellectually and physically deal with the environment through better use of time for improved decision-making. Furthermore, military personnel must

be able to exploit this use of time to improve their capacity to adapt through reorganizing, reequipping, rethinking, and reskilling.

2. The battle of signatures. Every item of military equipment possesses a signature, be it visual, aural, or electromagnetic. Larger military organizations also possess signatures. These include patterns of operations and exercise schedules as well as the indicators for impending military activity. The core idea for future military organizations is to minimize their tactical to strategic signatures, use recorded signatures to deceive, and be able to detect and exploit adversary signatures—across all the domains in which humans compete and fight.

3. New forms of mass. New approaches to mass manufacturing using 3D printing, and the ubiquity of autonomous systems across the land, sea, air, and space domains, herald a new era of warfare and competition. It involves large-scale conventional forces, the massed use of autonomous systems, and the wide-scale use of the tools of influence, including sophisticated algorithms. The core idea for future military organizations is to build forces with the right balance of expensive platforms and cheaper, smaller autonomous systems that will be more adaptable to different missions and be more widely available. This new balanced force must be employed using new twenty-first-century warfighting concepts and strategies by people whose training and education feature the integrated application of human and machine capabilities. We are returning to an era of mass warfare, but it must be developed and wielded in much more clever ways than before.

4. More integrated thinking and action. Over the past century, the domains in which humans compete and fight have expanded. Air, space, and cyberspace have joined the age-old domains of land and sea conflict. Unlike the counterinsurgencies of the past two decades, future military institutions must be able to operate in all domains concurrently. The core idea for future military organizations is to be part of an integration of all domains in a new, twenty-first-century national security operating model.

5. Human-machine integration. Robotic systems, big data, high-performance computing, and algorithms are already being absorbed into military

organizations in large numbers. The core idea for future military organizations is to augment human physical and cognitive capabilities to generate greater mass, more lethal deterrent capabilities, more rapid decision-making, more effective integration, more efficient training and education, and better experimentation and testing. New autonomous systems (robots) and algorithms (AI) will not just be tools used by humans. In many instances, they will be full partners with human beings in the conduct of military missions.

6. The evolving fight for influence. War has always been an intricate balance of physical and moral forces. Disruptive twenty-first-century technologies have not only enhanced the lethality of military forces at greater distances, but they also now provide the technological means to target and influence various populations (enemy and friendly) in a way that has not been possible before. The core idea for future military organizations is to ensure they, as part of a national approach, can improve their ability to influence adversaries, their leaders, and, where necessary, foreign populations.

7. Greater sovereign resilience. While deglobalization and "reshoring" of manufacturing ability has been under way for some years, the COVID-19 pandemic brought to the fore the requirement for greater national resilience in supply chains for critical materials and manufactures. This also holds true for societies and military organizations and their ability to deal with short-term crises (think natural disasters and pandemics) as well as longer-term, large-scale challenges such as the competition with China and climate change. The core idea for future military organizations is to mobilize people for large military and national challenges, while also developing secure sources of supply within national and alliance frameworks to ensure that supply chains cannot be a source of coercion by strategic competitors or potential adversaries.

## A New Appreciation of Time

Colin Gray has written that "every military plan at every level of war is ruled by the clock. Geographical distance, and terrain, translate inexorably into time that must elapse if they are to be crossed. On the virtual battlefield of cyberspace, electronic warfare is apt to mock geography and therefore time."[70] In war and competition, the clock is always ticking. The ability to exploit time is one of the most important considerations in the planning and execution of military activities.

In his study of time and warfare, *Fighting by Minutes*, Robert Leonhard proposes that four elements of time—duration, frequency, sequence, and opportunity—define the limits of political and military power.[71] Each of these elements has consequences in the preparation for, and conduct of, military activities. While all four are important, I propose that the two more vital elements in our exploration of future conflict and competition are *duration* and *frequency*.

Leonhard examines duration through the lens of events that have a start and a finish, even though these may not always be well defined. A range of variables might influence the duration of a conflict, including the size of respective forces, the importance of the goals sought, geography, and the level of training and technological sophistication of the involved combatants.

Over the past thirty years, Western nations have shown a desire to ensure that conflicts are more limited in duration. The desire for shorter wars is driven by the level of interests involved as well as the need to sustain political and public support for military actions. The reason why duration then becomes an important consideration in contemporary and future military activities is because competitors such as China and Russia have developed new modes of competing that play out over longer periods of time than we might otherwise prefer. The forms of political warfare and lawfare examined in the previous chapter often require durations of months, years, or even decades to yield results.

Confronting such methods will demand that Western nations adopt longer-term approaches to national security policy and strategy. Western nations will need to reconceive what they will accept in terms of duration in competition and conflict. In facing adversaries who might prefer longer confrontations, democratic societies will need to develop greater levels of strategic patience and resilience. Fortunately, this is not the first time that democracies have faced such a challenge. There are lessons for longer-duration competition from the Cold War that will be useful. But it will take time, resources, and political will to redefine how longer periods of time might be exploited to achieve strategic outcomes by various elements of national power in Western nations.

The second important element of time is frequency, or the pace at which things occur. Throughout military history, revolutionary change has generally occurred when one combatant is able to change the frequency at which it operates and thus to interfere with its adversary's frequency (and ability to respond).[72] Examples of exploiting a different frequency include Gen. William T. Sherman's march on Atlanta in 1864, the German army's invasion of France in 1940, and the U.S. Army's offensive operations in Kuwait to eject the Iraqi army in 1991.

In any future conception of military activity at the tactical, operational, or strategic levels, understanding frequency will be vital. Such an understanding must include how quickly events might occur or how many activities can occur concurrently or sequentially than we might be traditionally used to. Frequency or speed can be used to gain the initiative, to reduce an adversary's reaction options, and to impose paralyzing shock. The potential speed of future military operations is increasing, driven by new sensors and systems that allow commanders to more rapidly gain and exploit situational awareness. The physical speed of the tools of war is also increasing, especially hypersonic weapons and cyber operations.

The application of AI in all forms of information collection, analysis, dissemination, and decision-making will also influence how frequency in war might increase. The United States has a vision of what it calls *mosaic warfare*,[73] which seeks to more seamlessly stitch together the various kinetic and nonkinetic elements of military operations and better leverage information with manned and unmanned systems of capabilities. These will be capable of generating rapid speed and simultaneous operations that break down an adversary's facility for timely and effective decision-making. As one analysis of this method concludes, "Mosaic warfare places a premium on seeing battle as a complex system, using low-cost unmanned swarming formations alongside other electronic and cyber effects to overwhelm adversaries. The central idea is to be cheap, fast, lethal, flexible, and scalable."[74]

Similarly, Chinese documents and journal articles describe the *informatization* and *intelligentization* of warfare in the twenty-first century. The various ideas of Chinese scholars and military officers are associated with leveraging information to better connect various forces and generate a tempo across multiple military endeavors to paralyze an adversary and "break down their system." Chinese sources describe this *system destruction warfare* as the ability to paralyze the functions of an enemy's operational system, forcing an adversary to lose the will and ability to resist once their operational system cannot function. Paralysis can be generated through kinetic and nonkinetic attacks, as either type of attack may be able to destroy or degrade key aspects of the enemy's operational system.[75] While many of these approaches are not new, the Chinese concept sees them applying intelligentization to speed up the tempo of all military activities.

The intelligentization trend has the potential to become more striking in the future as advanced technologies such as robotics, artificial intelligence, biotechnologies, information warfare, cyber, and space capabilities proliferate. The use of these technologies will drive development of new warfighting concepts to cope with the speed of operations and the use of a combination of crewed and autonomous systems in the physical, cyber, and cognitive domains. The Chinese PLA, for example, has proposed concepts such as swarm attrition warfare, autonomous dormant assault warfare, and autonomous cognitive control warfare.[76]

An associated impact is that speed also deepens the strategic reach of military activities. This trend has become more obvious since the 1980s, and it has been accelerated by the reach of cyber operations and the influence generated by information war. If this is combined with the longer range and speed of tactical weapons systems, the challenges tactical commanders confront are compounded. This might be complicated even further by developments in biotechnology. In the longer term, the potential for biotechnologies to enhance the strength, speed, and cognitive abilities of humans will change the manner and speed at which humans fight.

Another consequence of increased speed in military activities is that it compresses the strategic-operational-tactical hierarchical framework. The consequence of this compression is that the potential rapidity of military activities also denies strategic military leaders, and national political leaders, the time to thoughtfully consider their options. The speed of certain weapons systems, particularly those that threaten strategic assets or whose payload is uncertain, will force the more rapid involvement of operational and strategic commanders in decision-making. It must drive better linkages between all sensors both in and outside the battlespace, as well as a demand for better decision support tools based on big data and artificial intelligence.

Speed will also have consequences for decision-making by strategic-level defense committees and strategic leaders. They may not always have the luxury of time or quality information to make decisions. As technologies evolve rapidly, weapons systems and equipment may become obsolescent faster than ever. While some of this challenge can be mitigated through new organizational design or operational concepts, eventually the weapon or equipment must be replaced. The long-term, multidecade focus that many Western defense procurement agencies currently have must change to be more agile and more time-conscious.

An increased frequency in military activities presents multiple challenges. Its compression of the strategic-operational-tactical framework necessitates that we must potentially rethink this entire approach to command and control and military chains of command. The potential for high-frequency military activity also means that many elements of extant military doctrine either are, or will rapidly become, obsolete. This will require every element of development military capability—including tactics, organizations, training, education, doctrine, leadership, equipment, and logistics–to transform and adapt to this new environment. It will place a premium on good leadership and the ability to change leadership styles and levels of control depending on the situation.

However, speed in military activities is a relative construct rather than an absolute. It is important only if it means operating at greater speed—or frequency—than an adversary. Additionally, higher-speed operations are not possible or even required for every endeavor. Achieving greater speed comes with trade-offs—faster platforms are more expensive and may also trade off performance in other areas such as sensor performance. Acting at the right time will always be more important than acting at speed. While some have imagined future conflict consisting of *hyperwar*, this will be neither achievable nor desirable in many circumstances.[77] No military institution can operate at maximum speed and capacity permanently. Rest, resupply, synchronization with other units and national assets, balancing kinetic and nonkinetic actions, and political considerations all have an impact on the tempo of military operations.

There will be very few rapid solutions should we be required to again conduct the types of missions recently undertaken in Iraq, Timor Leste, the Solomon Islands, South Sudan, and Afghanistan. Slowing down operational frequency over a longer duration may be preferred in these kinds of scenarios. Speed may not be necessary when dealing with the ideology and influence of terrorist and other extremist groups. Because of this, tempo is a much more relevant concept; it involves the element of speed that is used precisely when and where it is needed to generate advantage.

Therefore, understanding and being able to control tempo—the relative speed and rhythm of military operations over time compared to an adversary—is vital.[78] It is an important consideration in the command and control of military activities. It includes not only the frequency of operational activities but also the appropriate sequencing of operations and the capacity to transition between activities at the right speed. It also includes the ability to learn and adapt at a speed relatively faster than an adversary. However, controlling the tempo of operations is increasingly difficult because of the pace of change in technology and how this impacts the potential frequency of operational activities.

Robert Scales has written that "the one factor that will control the shape and character of a prospective conflict is time."[79] The twenty-first-century security environment will see time have an impact as new technologies appear at a brisk pace, and new weapons systems and AI will allow much more rapid tactical activities. Advances in technology that better link sensors with networks allow for a higher frequency of military operations. This creates an environment that is potentially more lethal and one where increased frequency means that humans will find it difficult to keep up.

For many of our military personnel and their leaders, this will be a bewildering security environment. Military institutions will need to provide their people with the ability to recognize change and adapt to it at a speed we have not seen before. And they must be able to do so at every level of war. War and competition are occurring with different durations and at a different frequency than we have seen in the past several decades. We must reconceptualize our understanding of time and its role in twenty-first-century conflict and competition.

## The Battle of Signatures

In the early 2000s, Adm. William Owens, USN, forecast that advances in surveillance technologies across multiple spectra would permit the U.S. military to deploy an unprecedented level of around-the-clock surveillance across the battlespace, giving commanders full visibility of events. Robert Scales described this as an "unblinking eye" of continuous sensor coverage.[80] Unfortunately, this idea would run aground on the shoals of operations in Iraq and Afghanistan. No level of persistent surveillance could predict the actions of low-technology insurgents in the urban or rural environments. No platform in the inventory, then or now, could see the intentions of a group of insurgents or the strategies of regional terrorist leaders.

Importantly, insurgents and terrorist groups (those that survived) learned to mask their various signatures to generate a level of stealth in how they operated, where and when they would attack, as well as their sources of support. We relearned over the last two decades that military operations are underpinned by a constant battle to collect information while hiding our information—especially physical and electromagnetic signatures. And despite predictions of an unblinking eye over the battlefield, the two-way nature of warfare almost guarantees that an adversary will counter any advance in the capacity to surveil their activities.

In the military, signatures refer to those ways that people, equipment, units, headquarters, networks, and other parts of the organization might be detected. They are the detectable characteristics of an object, including visual, acoustic, chemical, cyber, biological, and electromagnetic properties that derive from composition, structures, emissions, or surfaces. Different behaviors by military units are also detectable: communications, movement, formation, and effects on other entities can all be detected and analyzed. Even where individual components are well concealed or camouflaged, they may be compromised by their collective signature.

For a military institution to achieve a level of force protection and to continue operating against an adversary, it must possess a systemic approach to signature management. This includes dedicated activities to hide, minimize, or disguise friendly assets (in the real and virtual worlds) and disrupt, deceive, or distract threat sensors. The measures to achieve this may be material or behavioral, active or passive, and might require the application of different technologies and operational procedures. There is a long history of military institutions adopting such approaches. Historical examples include the Allied use of deception before the 1944 landings in France, Russian operations on the eastern front in World War II, and the Allied use of amphibious feints to deceive Iraqi forces in the lead-up to the 1991 liberation of Kuwait.

The groundbreaking inventions of the twentieth century such as radar, electronic warfare, stealth technologies, space-based sensors, and cyber systems all have an impact on the fight to find an enemy before they found you. This resulted in a constant struggle between lethality and dispersion, with dispersion being one of the best methods of lowering detection thresholds and making targets less attractive to attack if they were detected.

The Allied bomber campaign in Europe during World War II provides an excellent example of the constant back-and-forth between adversaries seeking to gain advantage by managing their signatures. The Allies' bomber fleets played a constant and highly technological game of cat and mouse with the German air defense system. German attempts to detect incoming bomber raids were countered with diversionary raids. Increasing accuracy of bombing raids led the Germans to camouflage real factories and construct decoys to lure Allied bombs.

The signature battle extended into the electromagnetic spectrum. The Allies developed a method that used radio beams for aircraft navigation. In successive generations called Gee (March 1942), Oboe (December 1942), and H2S (January 1943), the new navigational devices enhanced the ability of bombers to hit their targets at night. At the same time, the Germans developed their Wurzburg radar to better detect bombers and attack them using night fighters and anti-aircraft artillery. The Allies responded with the development of Window, the dropping of long strips of aluminum foil (chaff) to confuse German radars. Its first operational use in late July 1943, during British and American bombing raids on Hamburg, completely blinded German defenses. Eventually, the Germans developed their SN2 airborne radar set, which could penetrate Window and vector German night fighters toward Allied bombers.[81]

This cycle of developing and deploying capabilities, countering them, and then deploying counters to the counters continued throughout the war in the air over Europe. But the twenty-first century portends even more momentous breakthroughs in the capacity of military organizations to detect an adversary's various signatures and use this to rapidly attack it. Because of this, modern aircraft have systems such as infrared suppression systems to reduce the ability of infrared guided weapons to attack them. Submarines use a range of simple and advanced technologies (such as silencing equipment, hull design, and anechoic tiles) to lower their acoustic signature, which reduces the capability of antisubmarine forces to detect them. And on the ground, land forces use vehicle and individual camouflage, noise reduction, and electronic warfare to diminish their signature.

We should expect even more advanced systems in the future. With the advances in biotechnology noted in the previous chapter, it might become possible to turn plants into biological sensors. Imagine being able to use entire forests as sensing capabilities for military operations. Quantum, laser, and seismic measuring technologies also offer potential advances in sensing capabilities for military organizations as well as strategic surveillance.

The technological sophistication of potential adversaries, their mass, and their presence in every domain of war mean that the battle of signatures will be one of the defining aspects of warfare in the twenty-first century. Those institutions that can collect information on the various signatures of military organizations and turn this information into timely, actionable products will possess a decisive advantage. The new form of advantage will be constructed upon new, more robust, and agile communications networks and enabled by bespoke algorithms that can access data and increase computational power. It will be vital in the tactical environment where aircraft, missiles, ships, submarines, soldiers, and their vehicles will all live and die by their capacity to detect and kill their adversary while avoiding detection by that adversary.

Robotics and artificial intelligence offer another pathway to revolutionary advances in the collection, analysis, and rapid dissemination of tactical- and operational-level information and intelligence. Large swarms of different sensor systems—on the ground, in the air, and at sea—may provide military commanders with previously unimagined levels of surveillance and reconnaissance abilities. Matching the collection from the swarms with rapid collation, analysis, and dissemination via bespoke AI may provide a winning combination. Supplementing these swarms will be vast constellations of space-based collection systems that can collect many different forms of information across the visual and electromagnetic spectra.

Tactical forces will require new masking techniques to minimize their potential of detection. There is no chance of eliminating that potential entirely given the wide-scale proliferation of sensors. But new forms of camouflage, vehicle shaping, cyber warfare, and interference will need to be developed to counter the new sensor technologies. And this must be accompanied by a reinvigorated desire by Western military leaders to deceive adversaries actively and consistently. As an article in *The Economist* has noted, "Although countries continue to spy, propagandize, and sabotage, military deception . . . appears to be declining. Modern war is a profession, waged by complex machines and officers capable of wielding them. By contrast, deception is closer to an artistic enterprise."[82]

Another area requiring investment for signature management is strategic capabilities at home bases, military headquarters, and other strategic infrastructure. For example, if a high-readiness parachute brigade in the United States is about to deploy, a range of detectable signatures—aircraft movements, base lockdowns, and changes in the normal base patterns of life—will be evident. Likewise, we can detect changes in the pattern of life at strategic headquarters through parking lot occupation and even upticks in food deliveries. Online, changes in traffic across the Internet (given that many military institutions rely on an Internet backbone) can indicate potential military activities.

The implication is that signature management in the twenty-first century will be harder than ever but will be an essential military capability. Signature management must be conducted with a holistic approach that embraces the entirety of military activities from the tactical through the strategic levels. Signatures of all kinds at every level of military endeavor must always be measured and managed. The application of many sensors and AI analytical tools will be needed to protect friendly capabilities by reducing their signature and deceiving adversaries about our intentions. To master this art, military leaders must also become comfortable in merging the technological aspects of war with its more artistic, creative approaches that underpin military deception.

Signature management and military deception will be critical undertakings in the strategic competition and warfare of the twenty-first century. The battle of signatures will be a long-term endeavor necessitating significant investment across all the domains in which military activities occur. Of course, military institutions might choose not to do so. These will be the institutions that will fail. And in the decades ahead, they may not get a second chance to recover, learn, and recommence operations.

## A New Form of Mass

At the end of the Cold War, the United States and the Soviet Union deployed millions of regular military personnel in their global competition of ideology and military might. Combined, they possessed more than 17,000 nuclear warheads on land- and submarine-based intercontinental ballistic missiles and on various air- and ground-launched weapons systems.[83] In central and northern Europe, nearly one million NATO soldiers and 11,000 tanks faced off against Warsaw Pact forces comprising 1.1 million soldiers and 29,000 tanks.[84]

The fall of the Berlin Wall changed everything. Over the course of the 1990s, drastic reductions occurred in military forces in Russia, Europe, and the United States. NATO ground forces in central Europe were reduced by two-thirds to just over 300,000 in 2016. The United States, which had more than 5,000 main battle tanks in Germany in 1990, had none stationed there by 2014.[85]

The wars of the late 1990s and early 2000s have featured smaller but exquisitely trained professional military organizations from NATO countries and beyond. These forces have deployed outside their traditional areas of interest to undertake counterinsurgency and stabilization missions. Examples of this in the 1990s included the stabilization missions in the former Yugoslavia, East Timor, and Africa. In the early 2000s, these "exquisite and small" conventional and special forces were employed in Iraq, Afghanistan, and beyond.

But the wheel has turned again. Two enormous and vastly rich superpowers now confront each other in a battle of ideology, trade, and technology. The world has returned to an era in which a potential war might feature the clash of very large military organizations that have global reach. The United States and China are the two largest and most capable military organizations on the planet. Their combined annual defense spending approaches U.S. $1 trillion.[86]

In this new strategic competition, the mindset of mass clearly still plays a significant role in U.S. and Chinese thinking. While there has been a reduction in size as its quality improves, the PLA still has more than 2 million active military personnel. At the same time, the United States maintains 1.3 million personnel in its armed forces, and the Russians 900,000, with another 2 million reserves.[87] While the United States could in the past leverage better technology in confronting a larger Soviet conventional force, that is no longer the case. The Chinese armed forces have leapt forward in quality in the past decade, with many of their technologies achieving close to parity with the best U.S. weapons and sensor systems.

The return to large-scale military organizations and operations in the twenty-first century will not replicate the mass approaches of previous eras, however. Improvements in precision sensor and strike systems since the 1980s make massed formations much more vulnerable to detection and attack. Military units will therefore adopt more dispersed approaches that mask or delay the detection of the signatures of their individual platforms and collective units. This will occur across all the different physical and virtual environments.

Another characteristic of the new type of mass in military affairs is the balance of crewed and uncrewed systems. In the twenty-first century, the large number of troops, aircraft, ships, and intercontinental ballistic missiles possessed by military organizations will increasingly comprise a smaller proportion of their total strength. Widespread use of autonomous systems—on the ground, in the air, in space, and at sea—will provide a magnifying effect for human-centric units in both quantity and quality. Supercharging their impact, the mass of unmanned systems and AI will be (to use the Chinese term) informationized and operating on more secure networks and with greater autonomy than current systems. As T. X. Hammes has argued, warfare in the twenty-first century will be a question of "few and exquisite versus small, smart, *and many*."[88]

The desired effect of this new approach to mass military forces is not just to cover large areas of terrain in the physical world. These mass forces will possess the persistency, size, survivability, and precision to place nearly every large, exquisite military institution at risk. In the virtual world, massed systems will concurrently generate influence and other effects to shape the perceptions of national leaders and the populations of nations. Over the past two decades, the military organizations of Western nations have engaged in operations to win the "hearts and minds" of local people. In the coming decades, a strategic endeavor with global reach using big data, advanced algorithms, and clever messaging will seek to influence entire national populations.

The Chinese People's Liberation Army is taking this use of massed manned and unmanned systems to another level intellectually. While it has used informatization as a foundational doctrine over the last few years, it is exploring radically new operational concepts that exploit swarms of autonomous systems and AI-supported command and control. The authors of a 2020 article in the official PLA journal *Military Forum* describe a transformation of warfare where "the distinction between humans and weaponry becomes blurred, and it is even difficult to distinguish whether humans or machines are functioning." They describe new concepts such as autonomous cluster consumption warfare, autonomous latent assault warfare, and autonomous cognitive control warfare.[89] While these concepts might be untested in the real world, they at least represent new thinking that is not as obvious in the military institutions of the West.

In this competitive environment, where the best ideas must be applied to the large-scale use of military power, Western military institutions will need to undertake a wholesale redesign of their fighting concepts and organizations. They must be redesigned for an environment where the ratios of humans to unmanned systems will potentially be 1:100, 1:1,000, or greater. Unmanned systems in the 1990s and early 2000s were expensive or large, remotely controlled devices. Consequently, the battlespace was overwhelmingly human with a small proportion of unmanned capabilities. In the very near future, the size and cost of such autonomous systems will continue to decrease, which will heighten the attractiveness of their deployment.

Large and expensive systems may well still be needed, but they must be procured in fewer numbers and balanced with large numbers of small, networked, and inexpensive systems. For example, instead of buying a fleet of expensive nuclear or conventional attack submarines, a navy might buy only two-thirds of their requirement in the large, traditional platforms and invest the remainder of funds in hundreds of small, autonomous underwater vehicles. This might include a mix of different types including intelligent sea mines, hunter-killer systems, and vehicles for hydrography and interruption of undersea communications. These might then be used in declared "kill boxes" for a deterrence effect or simply used in more traditional ways to collect information or wait in stasis to attack larger and more expensive systems. Given the huge areas that many small and mid-size navies must cover, the logic to this approach is inescapable. This reasoning also applies for air- and land-based systems. The shift to this thinking has begun, with increasing experimentation and deployment of new uncrewed systems in the military organizations of China, Russia, and the United States.

There will be a huge first mover advantage for those who might reconceive the meaning of mass in twenty-first-century warfare. Being able to undertake surveillance across large areas of the earth, space, and cyber space and then exploit this with physical and nonkinetic capabilities in near real time with large swarms of small, strategically deployable autonomous systems must be the aspiration of any military institution that wishes to remain relevant in the coming decades. It will be impossible to do this with the "few and exquisite" approach to equipment and personnel that the West has developed since the end of the Cold War. The demands of future conflict will necessitate a shift in investment, thinking, and culture to adapt to this new situation. And it will require military institutions to be vastly more adaptable.

The current surge of technological innovation will demand that new-age massed forces and capabilities constantly evolve and adapt their strategies, organizations, and concepts of military employment. We are entering a new era of warfare and competition that involves large-scale conventional forces, massed use of autonomous systems, and wide-scale use of the tools of influence, including sophisticated algorithms and data analytics.

### More Integrated Thinking and Action

In the wake of World War I, strategy broadened from a military activity to a national one, including industry and other elements of national power. Since the end of World War II, there has been a shift to a more unified approach to military operations through the development of joint concepts and institutions. The various drivers for this include operational necessity (especially for smaller military institutions such as those of Australia or the United Kingdom) and political imperatives (such as the 1986 Goldwater-Nichols Act in the United States) that fundamentally redeveloped the structure and operation of the Department of Defense. The successes of the U.S. military in its 1991 and early 2003 operations against Iraq were testament to the benefits of the joint approach to military operations.

While the 1990s saw some conceptual development of jointness, the next major impetus came with the operations that followed the 9/11 attacks on the United States. Joint operations with their established interaction of army, navy, marine, and air force elements were no longer sufficient to address the requirements of contemporary warfare. A new domain—cyber—needed to be incorporated into operations. At the same time, existing mechanisms of government—diplomacy and economics among them—needed to be better integrated into the overall scheme of national strategy that addressed the threat of al Qaeda, ISIS, and other nonstate actors.

Over the past two decades, many nations have adopted a more integrated cross-government approach to national security affairs. More cohesive national approaches, rather than stand-alone military or economic methods, have been needed to address complex, longer-term, and multinational threats from terrorist groups and reemerging strategic actors. In the United Kingdom, this has been called the "joined up" approach, and it originated with the Tony Blair government in the 1990s. Over the past two decades, many nations have adopted this more joint, interagency, and multilateral approach. The result has been a deepening of understanding and relationships between different military services and civilian institutions involved in national security.

The new domains in warfare—especially cyber, information, and space—are increasingly central to military operations. They are becoming more integrated from the tactical to the strategic levels of operation. Closer integration featuring joint, coalition, and interagency activities demands the development of high levels of trust and collective planning.

The trend in twenty-first-century integration will continue, driven by the developments in geopolitics and technology explored in chapter one. The geopolitical competition across all elements of national power is a powerful driver for deeper integration of the various elements of warfare. The pace of change and the convergence of different technologies mean that surprise is more likely. More integrated information collection and sharing will be needed to mitigate this enhanced threat of surprise. Potential adversaries have designed concepts of military operations, such as China's systems destruction warfare, that focus on breaking down Western systems. These are clear warnings that military institutions and broader national security enterprises must develop even more tightly connected forms of planning and executing activities.

Democracies are naturally susceptible to Russian and Chinese approaches that exploit seams within national governments and ethnic groups and between governments and society. As Ross Babbage has proposed, "The nations of the West have suffered a lack of strategic integration and coherence over the past two decades. The strategic goals of the West in countering authoritarian states have been generalized, national leaderships have been distracted by counter-insurgency campaigns and internal politics, and the publics of the U.S. and other Western allies have been resistant to new foreign involvements."[90]

Russian and Chinese approaches also exploit the different mindsets, systems, and structures between nations such as the United States and its close allies. As Babbage has noted, "The defense and security systems of most Western and partner states are optimized for peacetime diplomacy and occasional deployment to conduct intense conventional warfare. Western electorates are fearful of triggering confrontation and the escalation of an argument."[91] This has allowed both Russian and Chinese strategists to apply a wide range of civilian, paramilitary, and military instruments to achieve their own strategic objectives beneath the threshold of violent conflict with the West.

Authors Qiao Liang and Wang Xiangsui have described the competitive environment from a Chinese perspective. Their work reflects a sophisticated understanding of the contemporary and likely future environment of strategic competition and warfare. In *Unrestricted Warfare*, Qiao and Wang describe a military institution with a different mindset from Western military organizations: one not dominated by the notions of the Western way of war that have proven successful until the dawn of the twenty-first century, a way of war that is showing increasing incapacity to be applied to twenty-first-century competition and warfare. They describe an environment where we must shift from "fighting the fight that fits one's weapons" to one where we "make the weapons to fit the fight."[92] The authors also demonstrate a sophisticated understanding of history—their own and Western. And in defining the future battlefield, they describe it as "everywhere":[93]

> Technology is again running ahead of the military thinking. While no military thinker has yet put forth an extremely wide-ranging concept of the battlefield, technology is doing its utmost to extend the contemporary battlefield to a degree that is virtually infinite: there are satellites in space, there are submarines under the water, there are ballistic missiles that can reach anyplace on the globe, and electronic countermeasures are even now being carried out in the invisible electromagnetic spectrum space. Even the last refuge of the human race—the inner world of the heart—cannot avoid the attacks of psychological warfare. All the prevailing concepts about the breadth, depth, and height of the operational space already appear to be old-fashioned and obsolete. In the wake of the expansion of mankind's imaginative powers and his ability to master technology, the battlespace is being stretched to its limits.[94]

The only way to address this stretching of the battlespace to its limits is for countries to develop more integrated, whole-of-nation strategies that allow them to apply all elements of their national resources. All arms of military power—on land, in the air, in space, on and under the ocean, and in the realm of information—must be integrated within a larger national security scheme.

These national approaches must be aligned within a more integrated Western alliance strategy to challenge the adventurism and advances of twenty-first-century techno-authoritarian states. It will demand a modification of existing bureaucratic structures, a different approach to risk, the building of an enhanced range of political warfare instruments, leadership, and the development of effective and inclusive strategies.[95] Whether the challenge is PLA systems destruction warfare or Russian strategies of limited action, deeper integration of institutions, bureaucratic systems, military organizations, state and federal authorities, and international partners is a key element of future strategy.[96]

At the tactical level, a sign of this deepening integration is provided by the development of joint tactical air controllers over the past two decades. The training of these people, normally corporals or sergeants (and their air force equivalents), now includes the capacity to call for bombers, fighters, naval gunfire, long-range rockets, artillery, and mortars in an integrated joint fires concept. This will extend to other areas, particularly in the cyber and influence domains. It will challenge the institutional cultures of many elements of military organizations. Some cherished functions may not exist in this environment, and the kind of people required may be very different than those we have traditionally recruited. Those institutions that can adapt most rapidly and overcome these cultural barriers will be more likely to succeed in twenty-first-century warfare.

Continuing integration will see the tactical-operational-strategic strata of the past five decades become more compressed. More junior people will undertake kinetic and nonkinetic actions that are more explicitly linked directly to political outcomes. While some might see this as an extension of the "thousand-mile screwdriver," it is in reality a more sophisticated method by which governments might use precise and measurable applications of military effects to achieve rapid political or strategic outcomes. This will stress extant military chains of command and bureaucratic systems. But we can and should anticipate this occurring—and evolve accordingly. As the authors of *Unrestricted Warfare* describe, "In a possible future war, the rules of victory will make extremely harsh demands on the victor. Not only will they, as in the past, demand that one knows thoroughly all the ingenious ways to contest for victory on the battlefield. Even more so, they will impose demands which will mean that most of the warriors will be inadequately prepared, or will feel as though they are in the dark: the war will be fought and won in a war beyond the battlefield; the struggle for victory will take place on a battlefield beyond the battlefield."[97]

Clausewitz wrote that "war is never an isolated act."[98] Integrating military, government department, and industry activities to achieve national purpose requires high levels of trust and extraordinarily competent collective planning. The continued development of more integrated approaches to strategy development and implementation is given additional impetus by the strong likelihood of states in the future seeking to gain strategic advantage in nonmilitary areas of power.

## Human-Machine Integration

Humans have always used tools to assist them in different types of work. These tools have often been adapted for use in warfare. The use of robotic systems and advanced algorithms is a continuation of this trend. The substantive difference in the twenty-first century is that there will be a greater level of autonomous activity to supplement physical activity and a significantly expanded use of artificial intelligence to improve and speed up human cognitive activities. Increasingly, this human-machine collaboration will extend into virtual worlds, which has been described as the *metaverse*.[99]

Robots and autonomous systems have already become commonplace in civil industry and the military institutions of many nations. The narrow range of military functions they presently undertake will expand over the coming years. For example, the majority of autonomous systems are currently employed in surveillance and information collection activities in the air, at sea, and on the ground. They have proved valuable in collecting information in operations in Iraq, Syria, Africa, Afghanistan, and beyond. But as their sophistication has grown, their potential applications (and appeal to military institutions) have increased.

Autonomous systems will proliferate in roles traditionally associated with equipment operated by humans. These include fighter and transport aircraft, armored vehicles, and logistic transport systems, as well as fighting and support ships at sea. There is some resistance to this in the military services, however. We might understand this if it was about the ethics of using autonomous systems. However, my observation over many years is that this is rarely the case. Largely, military institutions and their leaders seek conservative and reliable systems that they know will work. And they do not want anything to interfere with established incentive and recognition systems (such as promotions, appointments, and medals) in military organizations.

Autonomous robotic systems will collect information in the air, sea, land, and space domains and transmit it through networks that are secured and meshed through advanced algorithms. Information will be collated and analyzed—often without any human intervention—and then passed to humans and autonomous systems to either attack enemy targets, to influence individuals and groups, or to assist in other battlefield functions. Autonomous systems will also assume battlefield functions including clearance of mines, obstacles, and improvised explosive devices, bridge-building and road maintenance, logistic convoys, long-range strike, direct action activities traditionally associated with special forces, and aeromedical evacuation. These systems will also take over key functions such as air defense and counter-rocket and -mortar functions.

At the same time, counterautonomy systems will need to be developed and widely deployed. The aim of these systems is to prevent an adversary from exploiting the benefits of "small, brilliant, and many" autonomous vehicles. The capacity to undertake counterautonomy missions, however, is lagging that of autonomous systems in most military institutions. A 2020 Defense Science Board report noted a struggle "to find programs across the [Department of Defense] and whole-of-government with a counter autonomy focus."[100] As the benefits of fielding autonomous systems become more obvious to military institutions and nonstate actors, the development and deployment of counterautonomy systems and the accompanying doctrine, training, and education for personnel will become more urgent.

Farther away from the battlefield, we should expect a higher level of autonomy in the fighting and logistic ships that secure the seas and support expeditionary forces, as well as the aerial systems that undertake similar functions. Back at home, autonomous systems will be used for logistics and other support roles including the physical and cyber defense of military bases, airfields, and ports. Robotic systems will also be used as targets and potentially as trainers at training facilities. Almost all the technologies to make this happen are already with us, but they will need to be more affordable and more supportable with additive manufacturing. Importantly, they must be accompanied by new ideas about how joint combat units are trained, how they fight, and how expeditionary military forces are raised, trained, administered, career-managed, and sustained in their home locations.

Artificial intelligence and its associated technologies of big data, high-performance computing, quantum technology, and new-age encryption also promise to transform national security affairs and all forms of military endeavor. Rapid technological advances in these fields provide an opportunity for military institutions to rethink the conduct of planning, information gathering and analysis, cyber security, training, education, personnel management, logistics, and strategy development for competition and war.

The increased speed of tactical operations discussed earlier in this chapter must drive a rethink in military hierarchies, decision-making, delegation, and risk-taking. Traditional multilevel hierarchies in military organizations have historically provided planning and command support at different levels, while also ensuring that the span of command of individual leaders is not beyond what they can cognitively deal with. The application of AI and its accelerating influence in war means that there must be wholesale reexamination of current chain of command approaches and of the degree of delegation provided to subordinate commanders. Strategic and operational leaders will need to accept more risk, but at least this risk might be better informed by AI-assisted cognition and decision-making tools.

Another key driver for the use of AI in warfare is the convergence of large numbers of advanced sensors, extensive communication links, and an ever-increasing flow of information. As the quantity of information continues to increase, the capacity of humans to deal with it is not keeping pace.

The slowest node in military decision-making is becoming the human decision-maker. This does not mean humans should be excised from the process. They should be present where they play a meaningful, ethical, and creative role.

At the tactical level, artificial intelligence may change the balance of tactical military power. Kenneth Payne has described how it will "change the utility of force by enhancing lethality and reducing risk to societies possessing AI-warfighting systems. . . . A marginal technological advantage in AI is likely to have a disproportionate effect on the battlefield."[101] AI also has a range of possibilities for decision support, rapid simulation of the outcomes of multiple options, and better aligning tactical with higher aims. Different AI that uses human-in-the-loop and human-on-the-loop systems will support faster decision-making in an environment of rapid change. New forms of AI are probable that might support the more effective integration of joint and coalition capabilities.[102]

AI will also have wide application in the development of strategy. In the coming decades, AI will augment human cognitive functions, not replace them. But it will potentially have a greater impact in strategy than tactics. Military strategy is crafted within the policy framework of civilian governments. Senior officers engaging in development of strategy must appreciate the civil-military dynamic of this milieu. Their ultimate responsibility is to provide multiple genuine options expressed in a strategic context that explains how and why resources requested will solve or mitigate problems and assist to achieve policy objectives at acceptable levels of risk.[103]

There are multiple ways in which AI is likely to impact strategy development and execution. First, it will provide additional analytical capacity to ensure that desired strategic objectives result in securing the political goals of the nation. Second, it will assist military leaders and institutions to ensure strategic goals are aligned with military force size and structure. Third, bespoke AI might access massive troves of data to support strategic planning and decision-making so nations can align strategic objectives with their mobilization of logistic infrastructure, technology, and industrial capacity.

Finally, AI could potentially enhance a nation's capability to better align the achievement of its objectives with those of allies and other security partners.[104] Training and educating future military leaders and planners in these kinds of AI will be vital.

The tactical and strategic levels of military endeavor are just two examples of how AI might be used in organizations. AI will increasingly assist in complex defense force structure and personnel planning activities. It might be applied by military institutions to construct and provide the evidence for a regular share of national treasure to meet major known or potential threats. Another application for AI at the policy level may be for military institutions to survey and access suitable technology and industrial resources to produce equipment and other capabilities. A third, but by no means final, application might be for military institutions to use AI in a range of personnel functions to ensure they retain access to suitable quantities of quality people for the range of tactical, operational, and strategic endeavors for which they are responsible.[105]

The use of AI across the tactical, operational, strategic, and policy arenas will have another impact. Traditionally, higher-level organizations within the defense establishments of most nations are a balance of civilian and military personnel. Civilian officials provide a key connective function between government and the military and lead in development of defense policy. They are supported in this by senior military officers. Military leaders, however, lead with military strategy, operations, and tactics, ably supported (in most cases) by civilian public servants. The back and forth of dialogue in the civil-military environment has a multitude of benefits, not the least of which is the diversity of ideas that emerge from the interaction of two different cultures. But AI may become a third element of this relationship. AI will support civilians, military officials, and civil-military teams with decision-making. To that end, in the twenty-first century we may shift from thinking about civil-military relations to civil-military-AI relations.

The speed, breadth of uses, and quality of autonomous systems will continue to improve. Their cost—at least for small, mass-produced systems—will probably continue to drop, making them more attractive to military organizations as well as other security providers and nonstate actors. However, this is of less concern than artificial intelligence. Ultimately, it might be possible to develop a form of artificial intelligence that replicates human intelligence. Experts describe this as general AI or artificial general intelligence. There is a wide variety of views on whether this is possible. In a survey of expert opinions on the likelihood of this possibility, Max Tegmark proposes that it might occur in decades, potentially within a century, and perhaps not at all.[106]

The proliferation of AI and autonomous robotic systems in military institutions is not without risk. A 2020 report from RAND described three types of risks inherent in the application of AI to military equipment and military operations. The first is ethical; this includes issues such as accountability, human dignity, human rights, and the laws of armed conflict. The second risk area is operational; these risks manifest as trust and reliability, hacking and assurance, accidents, and other emergent risks. The final risk is strategic, which includes issues such as escalation management, proliferation, and strategic stability.[107] No nation has yet resolved these risks. But they must be addressed in parallel with technological developments. The ethical dimension is particularly important. We should ensure that these advanced technologies are designed and used by humans in accordance with respective national and institutional values.

For me, the key discussion is less about if it is *possible* for machines to replicate human intelligence than whether it is *desirable* for us to pursue this outcome. Despite the frailties in human decision-making, at least we own our decisions and are responsible for them. Intelligent nonbiological entities that truly lack "skin in the game" are unlikely to appreciate the full human implications of decisions in military operations. There is no guarantee that this form of intelligence would even listen to humans. It would essentially represent humans handing over all their key decisions to an alien race that we do not understand.

Amir Husain and John Allen describe this environment as hyperwar, "a type of conflict where human decision-making is entirely absent from the observe-orient-decide-act loop."[108] In some limited functions, this might be desirable: for example, a close-in weapons system that defends against high-speed missiles and artillery or defensive systems on aircraft.

High-speed analysis of options in decision-making would also be useful. But beyond this, the absence of humans in decision-making should be profoundly troubling for military and civilian leaders, as well as all members of our society. I am no Luddite, but there are so many unknowns with becoming fully reliant on AI that I can think of no human endeavor that would benefit from it—especially war.

### The Evolving Fight for Influence

Emily Goldman has noted that given the proliferation of nonmilitary activities such as cyber, information, and political warfare, "the strategic space below the threshold of war is now as strategically consequential as that above the threshold."[109] Perhaps no activity beneath the threshold of violence is as consequential as the ability to exploit information and influence individuals, organizations, and even countries.

The last two decades have seen a surge of interest in the conduct of information operations. Broadly, these are military activities that synchronize electronic warfare, cyber operations, human intelligence, social media, strategic communications, operational security, and other information capabilities to influence, disrupt, or corrupt the decision-making of competitors, enemies, and potential adversaries. At the same time, military organizations must protect their own information and decision-making and align their information operations with overall national approaches.[110]

The information domain is one of the most affordable in which to operate for state and nonstate actors. It is also difficult to attribute information operations to specific entities. As Sarah Kreps noted in a 2020 report on information operations, "Wielded well, these tools can generate massive payoffs for very little cost. With Russia, China, and Iran all seeking greater influence online, the dynamic somewhat resembles a Cold War arms race, but with information rather than missiles."[111] In other words, using information to influence actors and populations has low barriers to entry and offers potential adversaries a way to use a cost imposition strategy against the nations and military institutions of the West.

The future use of military forces, driven by autonomous systems, weaponized algorithms, and hypersonic weapons, highlights the potential for a more destructive form of warfare in the twenty-first century. However, as the Russians and the Chinese have demonstrated over the past decade, if objectives can be achieved without violence, most actors will do so. New ways of using information operations, lawfare, and deniable military and paramilitary activities offer different pathways to achieve strategic outcomes for state and nonstate actors.

Attempting to influence the mindset of an enemy commander and an enemy political leadership is as old as warfare. Methods include disinformation, subterfuge, and military deception; the history of war is full of examples. However, the means to influence leaders, their advisors, and the inhabitants of their nations are more sophisticated, more targeted, and more pervasive than ever before. The contemporary ability of states, corporations, and nonstate actors to devise and test strategic messaging—targeting different groups with different narratives—through the Internet and social media, adjust it, and access millions of users almost instantly is unprecedented. When strategic information campaigns are combined with other arms of national influence and coercion (economic, technological, legal, and paralegal, paramilitary, media manipulation, and subversion), they can be highly effective.

And they are effective. As one study by Massachusetts Institute of Technology researchers has found, social media has enabled false news to travel faster and penetrate further than true stories. Analyzing major news stories over a 10-year period, including 126,000 stories and 3 million tweets, the study found that false information outperforms true information.[112] Individuals and institutions are vulnerable to disinformation campaigns, and nations such as Russia and China have invested significant resources in their capacity to influence different individuals and populations.

China and Russia currently demonstrate an advanced capability to undertake strategic influence operations. To win without fighting is an old cliché but is the most powerful and apt description of their respective strategies. In his detailed study of Russian and Chinese methods, Ross Babbage has described an environment where their long-term strategies accept a high level of risk, are coordinated centrally, and adapt according to target reactions.[113] War and competition in the twenty-first century will see an increasingly sophisticated approach to influence operations—at every level—and an increasing quantity of them.

Dave Kilcullen has described these Chinese and Russian approaches as *liminal warfare*. It is a method that embraces ambiguity, where maneuver is never truly overt or clandestine. They "ride the edge, surfing the threshold of detectability"—tactics that have been honed over the past decade in reaction to the West's conventional force-on-force military dominance.[114]

Compounding the challenge, the primary strategic competitor for the United States—China—is also competing for global influence. It can apply a level of resources to its strategic competition with the United States in a way the Soviet Union never could. These technological approaches are supplemented by new forms of influence and coercion and are at the forefront of both Russian and Chinese strategies.[115]

A 2019 article in the Chinese *Military Forum* acknowledges that this type of conflict is inevitable and then notes that "traditional violent actions will evolve into hidden strikes, soft killing and consciousness control. Intelligent warfare will redraw the borders of war, but the standard for winning wars is still to achieve political gains."[116] Violence remains a key part of warfare. However, the wide availability of sophisticated and cheap tools for influence means that states and nonstate actors will lean more heavily on this method of competition and conflict in the twenty-first century.

An increasing preference for influence operations does not, however, portend the end of conventional operations. Sean McFate has recently written that "preparing for conventional war is unicorn hunting."[117] This is perhaps an overstatement or a misunderstanding of what conventional operations are. Conventional capabilities essentially prepare our troops for six core functions: to shoot, to see, to communicate, to influence, to command, and to support. Every military function in and out of combat can be boiled down to these six functions. These can be combined in various ways depending on the situation.

Military institutions cannot dispense with these functions. Combat (or the threat of it) remains an essential military activity. It provides for tactical and operational outcomes that support military strategies. But the possession of highly capable, lethal, and deployable combat capabilities across the domains also provides a powerful deterrent effect for potential aggressors. In many situations, the best way to influence the will of the adversary is to kill them, sometimes in large enough numbers to generate not only tactical defeat but also strategic shock. Combat, including close combat, will remain a vital element of nations achieving their desired objectives.

Every war and strategic competition is a finely tuned balance of violence and influence. At times, a high level of violence accompanies influence operations in war. At other times, during normal periods of competition, there is limited (but never zero) violence at the same time as an ongoing level of influence campaigns and strategies.

Democratic nations have largely forgotten the art of strategic influence. It is a function that many, including the U.S. government, honed to a very fine art during the Cold War. Governments and national security institutions will need to reinvest in the capacity for consistent, ethical, targeted strategic messaging if they are to combat the broad array of influence and disinformation campaigns from countries such as Russia, Iran, and China. This is not without cultural or legal challenges. As Thomas Rid has written in his masterful study of disinformation, *Active Measures*, "It is impossible to excel at disinformation and democracy at the same time."[118]

The influence and disinformation campaigns undertaken by authoritarian regimes can be corrosive to democratic societies and their institutions. Therefore, democratic nations must also be able to counter the disinformation campaigns of adversaries. This requires an integrated approach across government, including military institutions. Democracies have been slower than their authoritarian competitors to recognize the challenges and opportunities of influence campaigns. In the coming decade, they must catch up through comprehensive programs that build greater societal resilience against the disinformation for digitally enabled authoritarian regimes.

Military institutions must also invest more resources in generating influence. Building this capacity may necessitate military organizations divesting themselves of legacy systems of decreasing utility in war. It will demand they invest more people, equipment, relationships, and resources in the capacity to influence the mindset of an adversary. Additionally, enhancing this capability will require a reexamination of the legal authorities that military organizations possess to conduct such activities.

Russia's *National Security Strategy 2020* notes that "a 'global information struggle' is now intensifying."[119] Military organizations must invest in influence operations at multiple levels. Leveraging information and the fight for influence must be integrated within a national approach, be undertaken with a patient, long-term mindset, and be highly adaptive. Western military institutions, and the national security communities they collaborate with, must become the masters of this approach. As a British doctrinal publication on information activities notes, "We have reached the tipping point. Information is no longer just an enabler, it is a fully-fledged national lever of power, a critical enabler to understanding, decision-making and tempo, and a 'weapon' to be used from strategic to tactical level for advantage."[120]

The Western way of war, dominant over the last century and focused on kinetic action, is weary but not dead. It can and must evolve to meet the challenges of twenty-first-century warfare. Military institutions must evolve along with the increased speed and reach of the Internet and social media. They must invest in the people, ideas, and organizations to be able to undertake more holistic and widespread influence activities against competitors and potential adversaries.

### Growing Sovereign Resilience

In December 2020 the Australian federal parliament's Joint Standing Committee on Foreign Affairs, Defence, and Trade released a report on the implications of the COVID-19 pandemic for Australian defense and national security affairs. The most important finding of the report was that "COVID-19 exposed structural vulnerabilities in some of the critical national systems that enable Australia to function as a secure, prosperous first-world nation. Many of these vulnerabilities are caused by supply chains that rely on just-in-time supply from the global market. In some cases, this is exacerbated by supply coming—in whole or substantial part—from companies that are subject to extrajudicial or coercive direction from some foreign governments."[121] While specific to Australian circumstances, the report offers observations and recommendations relevant to many other nations.

The growth of a more globalized economy has led to a security environment where the old benefits of geography no long guarantee national security or sovereignty. That is not to say that geography does not matter; it does. It still complicates defense strategies and ensures that aggressors must think carefully about their ability to project and sustain military forces. However, as Stephan Fruehling wrote in 2017, "The loss of security from distance means we need to consider battle damage to Australian infrastructure and industry itself."[122]

The events of 2020 and 2021 have brought to the fore the importance of national sovereignty in critical manufacturing capabilities as well as critical national systems. As one commentator noted, "Globalization was already heading in reverse before the crisis and this will only be reinforced by the virus experience. Businesses will rethink long and complex supply chains, governments will feel compelled to ensure domestic capacity in more areas deemed critical to the national interest."[123]

Many nations must now examine what domestic capacity is needed and decide the costs and benefits of bringing home the production capacity to do so. This will have a material cost. But it also means that national resilience henceforth will be a more central aspect of developing strategies for national security. Nations are rarely self-reliant in many vital commodities and man-ufactured goods, so a reliance on external supply chains has become a more important element in national security planning. Nations do not need to physically threaten or attack a nation to coerce them; threatening their sup-ply chains or imports has become a potential strategy for many nations.

The Chinese Communist Party has increasingly resorted to using economic coercion.[124] Its aggressive "Wolf Warrior" diplomats, so named after the pop-ular Chinese movie of the same name, have looked to coerce nations across the globe that China perceives as threatening it in some form.[125] Its behav-ior in the wake of the COVID-19 pandemic, where it threatened nations in Asia and Europe, is instructive. We should expect that this behavior will continue and potentially expand in the coming decade. Therefore, the ques-tion arises of how nations might keep a level of sovereignty over their supply chains, particularly those related to military affairs.

There are very few nations capable of achieving a full measure of sovereignty when it comes to national defense. The United States and China are perhaps the only nations that might achieve self-reliance over the key industries and manufactured goods required for national defense capabilities. The rest of the world remains subject to some level of dependence on other countries for their capacity to defend themselves. But many nations will look for more secure ways to do this over the coming decades to reduce the threat of economic coercion and ensure access to key equipment, energy, and supply classes.

This is complicated by several factors. First, since the end of the Cold War, Western nations have reduced the level of their defense expenditures, resulting in many defense companies either consolidating or going out of business. Where large standing forces during the Cold War may have ensured viable defense industries in many nations, this is no longer the case.

Second, the process of globalization in the past thirty years has seen the offshoring of many industries to Asia. For important manufactured items and components, many nations are reliant on the global trading system rather than their own national workforce. While this is a resilient global system, some key high-demand items such as medical supplies suffered shortfalls in availability, as we have seen during the COVID-19 pandemic. This could also be the case with military commodities in the event of a major conflict.

A third complication is that many of the components of medicines, electronic goods, and fertilizers are now produced by the West's major strategic competitor—China. Beijing's deliberate strategy has been to become the world's factory. As one recent report notes, "China has also played by its own economic rules. Chinese manufacturers have undercut successive industries not only by outcompeting, but also by cheating. Chinese industry has thrived by stealing intellectual property on an unprecedented scale, and by benefiting from unequal investment practices imposed on foreign companies operating in China."[126]

A 2020 report by the London-based Henry Jackson Society found that nations such as Australia, Canada, and the United States had developed a "strategic dependency" on China. Across a range of services, manufacturing, and infrastructure capabilities, these nations have become reliant on Chinese capabilities—even though Australia, Canada, the United States, and the United Kingdom comprise a trading unit larger than other global trading blocs such as the European Union.[127] Such reliance has become a critical vulnerability for many nations and is increasingly recognized as such by governments and commercial entities.

Nations therefore must commence a long-term shift back to more self-reliance when it comes to their capacity to defend themselves and resist economic coercion. In some cases, this will mean that countries will choose several key aspects of defense production and capability that they do not wish to have produced offshore. Industries such as ship-building and vehicle manufacturing might be included in this approach. It should also incorporate research and development capabilities in areas such as artificial intelligence, cyber defense and offense, signature reduction of platforms and activities, and the supply of vital ammunition, precision munitions, and other consumables.[128]

Where nations lack the funding or the know-how to develop their own indigenous capabilities, they may form blocs of trusted nations to collectively develop and supply these goods and services. These blocs might also be used to collectively produce like items to reduce their costs through higher volumes. Three key blocs are likely to be based on the existing Five Eyes community (Canada, the United States, the United Kingdom, Australia, and New Zealand); the newly emerged Quad grouping of India, Japan, the United States, and Australia; and the European Union. Others will emerge, with the United States probably being a vital contributor, ally, or partner in each.

A reshoring of defense industries, driven by the new strategic competition and the aggressive, coercive behavior of authoritarian regimes, will need to accelerate over the coming decades. Nations will have to make difficult decisions about the indigenous capabilities that they cannot do without. But there will be a greater level of national sovereignty in supplies of military capabilities. It will shape warfare by ensuring nations own the capacity to sustain their military forces over longer periods of time. It will diversify supply chains, making them more resilient to external shocks.

An enhanced level of national resilience will also permit a more rapid mobilization of people in times of crisis. Mobilization might be needed for short-term crises such as natural disasters. It may also be required for longer-term challenges, which could demand a large proportion of the population being required to serve in military or other civil assistance roles in military or climate change–related emergencies. Importantly, building sovereign resilience to shocks is an important element of competition and conflict in the twenty-first century. It will add another element to a nation's ability to resist economic coercion and build capabilities to deter aggression.

## WAR AND COMPETITION: THE FUTURE

Sir Hew Strachan has written that "one of the central challenges confronting international relations today is that we do not really know what is a war and what is not."[129] This is because of the impacts of information and the ability of nations to exploit it in a way that has never been possible before. Our competitors have demonstrated an early understanding of this. Chinese military commanders are guided by the concept of "combat space is shrinking, war space is expanding," showing a keen appreciation of the strategic impact of information and influence.[130] This is an important element of how the character of war is changing. Military institutions in democratic nations will need to be highly adaptive in this changing environment if they are to generate advantage in future competition and war.

Military institutions might respond most effectively to these changes in the character of war in three areas. Considered in isolation, they provide insights into how military power is developed and applied. Combined, they are a powerful manifestation of a nation's investment in defending its territory, its values, and its interests. Unfortunately, they represent three areas that are subject to insufficient scrutiny or examination in many books on technology or future war.

The first area is ideas. Ideas provide the intellectual glue that coheres military units and institutions internally and binds them to the nations that they serve. Ideas extend from tactical concepts of employment all the way through to military strategies that nest within national grand strategies. Without the constant attention to developing and testing new and evolved ideas, military institutions do not just mark time; they doom their people to bad leadership and failures in military operations against those who have made the professional and intellectual investment in new, improved, and tested ideas.

The second area of impact is evolved and new military institutions. Modern military institutions were born in the wake of the establishment of the nation-state. The structure of the military institution has evolved over centuries, but its purpose has remained largely the same—to produce effective and sustainable fighting forces that can defend their nation and achieve missions set for them by their national leaders. In periods of significant change, military institutions have demonstrated a varying range of behaviors. Some have avoided change and retreated into tradition. Others have recognized change and opportunity early and undertaken the necessary intellectual and structural adaptation needed to remain effective. Those institutions that have nurtured learning cultures, invested in new ways of thinking and operating, and subjected these to rigorous testing have been more likely to succeed.

The final impact of twenty-first-century changes in war is people. If ideas are the glue that binds military units and institutions together, it is people who *are* these institutions. Good military institutions recognize the centrality of recruiting, training, and educating the very best people that they can access within their society. The approach to doing so has evolved over centuries, with most military institutions now possessing sophisticated ways of developing, leading, and managing their military personnel. But the challenges of the current era are straining many extant practices; military institutions will need to adapt their approach to personnel development if they are to retain some certainty of success in future military endeavors.

The following chapters explore these three elements of military capability. They will identify where change is required over the coming years in the different aspects of ideas, institutions, and people. In developing their approach to these three critical areas, Western nations may hone and significantly improve their ability to generate effective twenty-first-century military power.

# 3

## INSTITUTIONS, IDEAS, and FUTURE MILITARY EFFECTIVENESS

On a spring day in the late nineteenth century, the fishing schoo-
ner *Eliza Drum* departed port in Maine and sailed to its fishing
grounds off Newfoundland, where it joined other American fishing vessels. These ships went about their trade under the protective watch of
a U.S. warship, *Lennehaha*.

In the vicinity of the American fishing fleet and their naval overwatch
was the British warship, HMS *Dog Star*. The British man o' war enforced
a three-mile limit off the Canadian coast. Any American fishing vessel
that encroached on this limit would be boarded by the Royal Navy and
its catch seized.

On this particular spring day, *Eliza Drum*, in its exertions to maximize its
catch, had fished just inside the three-mile limit. Shortly afterward, sailors
from the Royal Navy warship boarded and seized the American fishing vessel. Events then deteriorated: soon after, *Lennehaha* and *Dog Star* were firing
upon each other with their bow guns. Thus began a war between England
and the United States. Unfortunately, the American military had been
underfunded and its Navy almost nonexistent. The U.S. government, under
pressure from the nation to respond to the Royal Navy attack on *Lennehaha*,
sought options in its new conflict with Britain. They found their solution in
contracting out the war to private enterprise.

Twenty-three "great capitalists" formed a syndicate and proposed to Congress to take charge of the war. The syndicate would "assume the entire control and expense of the war, and to effect a satisfactory peace within one year."[1] The U.S. government accepted the offer, and the syndicate rapidly turned to preparing the nation for war, largely ignoring politics and public opinion: "Politics were no more regarded in the work they had undertaken than they would have been in the purchase of land or of railroad iron. No manifestoes of motives and intentions were issued to the public." The syndicate developed a defensive strategy (defense by means of offense) and decreed that the war "must necessarily be quick and effective."[2]

The syndicate also designed and manufactured a range of new weapons including long-range guns, "motor bombs," and naval vessels called "crabs," which proceeded to defeat a fleet of Royal Navy warships in battle off the Isle of Wight. The Royal Navy was prepared for ordinary naval warfare, but the Americans "inaugurated another kind of naval warfare, for which it was not prepared."[3] Shortly afterward, the American syndicate destroyed elements of the city of Cardiff in a demonstration of a new weapon of mass destruction. The British government, unable to respond, negotiated for peace. Within a week, the American syndicate travelled to London and formed the Anglo-American syndicate of war.

Of course, there was no United States–England war at the end of the nineteenth century. American author Frank Stockton published *The Great War Syndicate* in 1889, a science fiction novel that envisioned how future military operations might be undertaken with many of the new technologies that had recently emerged. It is representative of the fascination with future conflict that was driven by technological and societal changes throughout the nineteenth century.

––––––

Humans have always been captivated by the future. This is no less true for generations of military leaders who have sought to anticipate future conflict to better prepare their forces. Today the exploration of future military endeavors is informed by a century of high-tech warfare, research and development, and wargaming activities. It is undertaken by highly skilled military officers, civil servants, scientists, and academics.

Advanced technologies and a new competitive security environment drive military institutions to experiment with different ideas about how military operations are conducted. The new ideas must be suitable for both competition and conflict and be designed to align with a more integrated approach to national security. At the same time, new and evolved institutions and organizational structures must be explored and established to provide more effective and efficient ways to wield military power.

The magnitude of the strategic competition ahead calls for significant investment in future planning to build creative solutions that provide strategic advantage for military organizations. To inform this future planning, we can apply the lessons of previous eras in which surges in innovation necessitated explorations of new military ideas and institutions. The Cold War and post–Cold War eras saw institutional and individual investment in the exploration of future war. Whether it was the large volume of reports from think tanks such as RAND or popular fiction that explored military operations, anticipating future forms of conflict was a constant feature of strategic thinking during these eras. These efforts and their results offer us important insights into how we might prepare for competition and conflict in the twenty-first century.

This chapter will explore the institutional and conceptual implications of continuity and change in war. Both continuity and change must be considered in building military institutions and ideas that will be effective and competitive in the decades ahead. The principal question is this: How do we ensure the competitiveness and effectiveness of military institutions, at all levels, within an integrated twenty-first-century national security approach?

Obviously, the ultimate determinant of military effectiveness is whether military organizations can deter conflict and, if they cannot, can fight and win battles, campaigns, and wars to meet political objectives. To do so in the twenty-first century demands imaginative approaches for how military organizations respond to the drivers of change in the strategic environment and how they think about military effectiveness.

## THINKING ABOUT FUTURE MILITARY EFFECTIVENESS

Contemporary approaches to examining future war began in the wake of the U.S. coalition victory over the Iraqi military in 1991. Witnessing the achievements of stealth aircraft, precision weapons, computer connectivity, global logistics, joint operations, and rapid ground maneuver, many defense analysts proposed that military institutions were at the dawn of a revolution in military affairs. The term gained wide usage in the military in the United States and other nations. Many viewed the 1991 Gulf War as the harbinger of revolutionary change in military capabilities and operations.[4]

These efforts built on earlier initiatives to ascertain the impact of computers and more precise missile technologies during the third Industrial Revolution. In the wake of the 1973 Arab-Israeli War, the U.S. Army had undertaken a detailed study of new technologies. The use of electronic warfare and anti-tank weapons and the increased lethality of infantry were of interest, given the likelihood of large-scale armored warfare in any war against the Warsaw Pact countries. U.S. Army studies of new ideas for warfare eventually led to doctrines such as active defense and AirLand Battle.

At almost the same time, Russian military leaders and theorists were examining the impact of new technologies on warfare. From the late 1970s, Soviet theorists and academics began writing extensively about a military technological revolution. As Andrew Krepinevich has noted, the Soviet focus was a result of their "anxiety of watching a more technologically advanced United States develop new technologies and move to incorporate them into new military systems."[5]

In late 1990 the head of the U.S. Department of Defense Office of Net Assessment, Andrew Marshall, commissioned a report on whether Soviet theorists were correct in their belief that technological advances would result in significant change to warfare.[6] Throughout the 1990s and early 2000s, military institutions, academia, and defense industry explored historical cases of military revolutions. Their hope was that this might inform conceptual, organizational, or technological advantage in future warfare.[7]

Today, a variety of nations, academic institutions, and military services are publishing assessments of advanced technologies and their impacts on future conflict. Many nations use these assessments to inform change in their organizations and concepts. However, the size, technological sophistication, and speed of U.S. and Chinese developments mean all other efforts are likely to be shaped in some form by them.

In the United States, a dominant idea about future military effectiveness over the past several years has been multidomain warfare. The idea has guided equipment development and conceptual exploration about future war in the U.S. military. Shaped by the 2018 *National Defense Strategy* and the 2021 *Interim National Security Strategic Guidance*, which acknowledges the return to long-term strategic competition with China and Russia, the U.S. military is now focused on integrating multiple elements of national power to build a highly lethal force within an interagency and alliance framework.[8]

The services of the U.S. Defense Department have adopted varying approaches to the development of this more lethal and effective multidomain force. The Army in December 2018 issued its capstone plan for multidomain operations, "The U.S. Army in Multidomain Operations 2028." It describes convergence of capabilities in the physical, cyber, and influence domains at multiple levels so that they might "prevail in competition" but also be able to "penetrate and disintegrate enemy anti-access systems and exploit free- dom of maneuver."[9] In 2016 the Air Force chief of staff listed multidomain operations as one of his three focal areas. Later, he noted that "multi-domain operations is about thinking through how we penetrate, where we need to penetrate; how we protect what we need to protect inside a contested space; how we persist in that environment for the period of time."[10]

The multidomain concept has begun to influence conceptual development and the procurement of weapons systems, and it is also shaping organiza- tional reform. The U.S. Air Force has created a new Space Force that over- sees U.S. access to space and the sustainment, protection, and generation of space-based capabilities. The U.S. military has continued to explore differ- ent elements of the multidomain theory. One example of this is the work by the Defense Advanced Research Projects Agency to develop mosaic warfare, described as "a new asymmetric advantage—one that imposes complexity on adversaries by harnessing the power of dynamic, coordinated, and highly autonomous composable systems."[11]

The foundational idea of mosaic warfare is to overcome Chinese and Russian advantages in precision, long-range attack systems (including kinetic, cyber, and influence activities). It also seeks to generate a sustained, long-term advantage for U.S. forces. One study of this concept by the Center for Strategic and Budgetary Assessments describes it as "leveraging distributed formations, dynamic composition and re-composition, reductions in electronic emissions, and counter-C2ISR [command, control, intelligence, surveillance, and reconnaissance] actions to increase the complexity and uncertainty an adversary would perceive regarding U.S. military operations and degrade the decision-making of opposing commanders."[12] While several nations are acknowledged as the potential adversaries driving this work, much of the U.S. effort is focused on what it sees at the principal threat to global security—China.

One of the best-known treatises on war, *The Art of War*, is of Chinese origin. Upon this heritage, the Chinese People's Liberation Army is constructing new theories of competition, warfighting, and military operations. China, and the motives of the Chinese Communist Party, are at times difficult for Western observers to understand. But there are sufficient Chinese documents available, such as defense white papers and the *Military Forum* journal, that supply insights into their thinking about how their institutions and ideas will underpin their future military activities.

American scholar Tai Ming Cheung, a long-time observer of Chinese military developments, has described how "China's leaders see science, technology, and innovation as essential ingredients in the pursuit of power, prosperity, and prestige. This is especially the case in the military realm."[13] So this has proved over the past two decades. With its growing military ability, the Chinese PLA has established new institutions to enhance its capacity to influence competitors and conduct long-range strike against military forces around the periphery of China. One such new organization is the PLA's Strategic Support Force.

Emerging from significant reforms to the PLA in 2015, the Strategic Support Force was established to command space, cyber, electronic, and psychological warfare capabilities. It brought together what had previously been disparate capabilities located in different land, sea, and air services under a unified and more strategic approach. Not only is this institution unique, it also is designed to implement a key idea from the 2000s, the *three warfares*—legal, public opinion, and psychological.[14]

At the same time, China has pursued new ideas to enhance its capacity in the global competition with the United States. Strategically, it seeks a more integrated approach to technology development under its civil-military integration program. China's national defense strategy, released in 2019, describes an approach called active defense whereby it sustains a defensive strategy as well as the capacity for offensive operations at the operational level.[15] Like the United States, China has embraced multidomain operations, but from its own perspective. Its National Defense Strategy notes that "efforts have been made to build the military strategy into a balanced and stable one for the new era, which focuses on defense and coordinates multiple domains."[16]

Three core ideas of this Chinese approach to future war and competition stand out. The first is the concept of gaining advantage through informatization and intelligentization of warfare. This refers to being able to bring together all forms of information for analysis, supporting this process with AI and other automated systems, and then disseminating this in a way that can be acted on more quickly and decisively than an adversary can. A second core idea is the theory of systems destruction warfare, which describes how future war will be "won by the belligerent that can disrupt, paralyze, or destroy the operational capability of the enemy's operational system. This will be achieved through kinetic and nonkinetic strikes against key points and nodes."[17] The final core idea is political warfare. This has been explored over the past several years but has been a prominent element of Chinese behavior in 2020 as it has sought to coerce, bully, and threaten multiple nations with economic sanctions.[18]

The Chinese Communist Party and the PLA possess massive conventional military forces that have been the beneficiary of new injections of highly advanced, lethal, and precise weapons. However, these forces are made significantly more capable by the large investment in the development of new institutions and ideas to wield them. The structural and conceptual reforms of the various arms of the People's Liberation Army over the past decade— and which we should expect to continue—will pose a serious and ongoing threat to Western military forces. Any underestimation of Chinese military ability in future competition or warfare is likely to have disastrous consequences for our military institutions and possibly our nations. We must significantly upgrade our capacity to understand not only what equipment the PLA possesses, but also how it thinks.

In exploring the efforts of these nations to divine optimum pathways for successful competition and conflict, several themes are apparent. The first theme is that military institutions have accepted that future warfare will be very different from what has preceded it—a recognition of the changing character of war. It is also an acknowledgment that new technologies and the new geopolitical competition have changed how nations compete and how they may fight against each other or fight through the use of proxies. As the British describe in their 2021 document, *Defence in a Competitive Age*, "The nature and distribution of global power is changing as we move towards a more competitive and multipolar world."[19]

A second theme is the desire of many nations to integrate military activities more tightly within national approaches to power. Whether it is the more efficient and effective use of strategic and operational capabilities or the desire to avoid systems destruction by Chinese forces, a tighter linkage of all military capabilities in a national power construct is a feature of current military development. It will shape all future developments in the institutions and ideas of Western military forces.

The third theme is that the major military powers appreciate the need for new concepts and organizations that are appropriate for twenty-first-century competition and warfighting. Demonstrations of this include the formation of the Chinese Strategic Support Force, the Australian Information Warfare Division, and the U.S. Space Force. At the tactical level, new multidomain task forces and Russian battalion tactical groups stand out as new organizations that can use new technologies and ways of collecting, sharing, and exploiting knowledge. At the same time, these new ideas about competition and conflict are an acknowledgment that tactics, processes, and strategies must be developed, implemented, and adapted more quickly than ever before. Speed and adaptive capacity at all levels must be principal features of every future military institution and idea.

Finally, notions of future warfare will continue to evolve as technology, strategy, and the global system change. Large nations such as the United

States and China will continue to anticipate change while concurrently responding to the actions and strategies of external actors. This environment places a premium on the ability to adapt at every level and at the appropriate pace. This notion of adaptive capacity will be explored later in this chapter.

These themes demonstrate just how difficult it is to anticipate the form and duration of future conflict. The challenge institutions face is to gain an understanding of what is changing (and at what pace) and what is not (continuities). They must then use this knowledge to construct achievable theories of victory and translate this into the resourcing to build organizations (including people and equipment) able to achieve these theories of victory.

Military institutions must make a series of bets on the likelihood of possible future scenarios eventuating and of their potential impacts. They must do this (at least in Western military institutions) in an environment of close government oversight and military budgets that are rarely assured or, if they are, rarely sufficient to address all potential future threats. The better these military institutions can understand the likely future security environment, potential changes and continuities, threats, and opportunities, the better informed their bets might be. To ensure they can make such investments with some level of assurance, military institutions need to evolve their current approaches to military effectiveness.

## FRAMEWORK FOR TWENTY-FIRST-CENTURY MILITARY EFFECTIVENESS

As we explored in the previous chapter, there are areas where warfare is likely to change as well as aspects where it is likely to remain unchanged. An appreciation of these changes and continuities is important because it provides a starting point for investment in new and evolved ideas and institutions. While this chapter is focused on the development of an evolved framework for military effectiveness, we should base this work on existing models.

In the late 1980s Allan Millett and Williamson Murray conducted a significant study of military effectiveness. Their study grew from a conference held at Harvard in 1980 on the military intelligence estimates before the two world wars of the twentieth century. Millett and Murray subsequently wrote to the director of net assessment in the Pentagon, Andrew Marshall, proposing a study that might identify elements of military effectiveness for the U.S. military. Marshall subsequently endorsed and funded the study.

The result of the study was a three-volume series of books on military effectiveness published in 1988. Covering World War I, the interwar period, and World War II, each volume used an analytical framework to explore different levels of effectiveness for multiple countries. These historical periods were chosen because they provided insights for the ongoing (at that time) strategic competition between the United States and the Soviet Union. Having entered another period of strategic competition, the analytical framework of Millett and Murray therefore has application (in an updated format) for the current environment and contemporary military institutions.

Another important study of military effectiveness was undertaken by Risa Brooks and Elizabeth Stanley in their 2007 book *Creating Military Power: The Sources of Military Effectiveness*. Brooks and Stanley proposed that the study of military effectiveness is necessary because it provides important insights into military power, a core concept of international relations. Importantly, the authors offer a model for examining military effectiveness that uses four attributes: integration, responsiveness, skill, and quality.[20]

Brooks and Stanley define *integration* as the degree to which different military activities are internally consistent and mutually reinforcing. *Responsiveness* is the capacity to tailor military endeavors to a nation's own activities, the capabilities of an adversary, and external constraints. *Skill* is about military personnel and how well they can execute tasks and carry out orders. Finally, *quality* measures the ability of a military to provide itself with highly capable and reliable weapons and equipment.[21] Each factor is essential for military effectiveness, and the four attributes are useful guides for our development of a twenty-first-century framework for military effectiveness.

Finally, in a 2004 study of military effectiveness, Stephen Biddle and Stephen Long examine why democratic nations are unusually successful in war. They proposed that success by military organizations in democracies is "a product of many democracies' superior human capital, civil-military relations, and cultural background."[22]

But what is military effectiveness? Millett and Murray define it as "the process by which armed forces convert resources into fighting power."[23] Brooks and Stanley define military effectiveness as "the capacity to create military power from a state's basic resources in wealth, technology, population, and human capital."[24] These are extraordinarily helpful definitions for our purposes here and provide an excellent starting point.

However, as we have already explored, war continues to evolve. Therefore, our notions of military effectiveness must also evolve. Twenty-first-century war and competition involve a more refined balance of violence and influence, the requirement to operate in all domains concurrently, and the need to better integrate military activities within a national approach. I therefore propose a slightly amended version of the definition of military effectiveness. In the twenty-first century, military effectiveness will be the process by which military forces convert resources into the capacity to influence and fight within an integrated national approach.

The framework Millett and Murray developed to assess military effectiveness focused on the political, strategic, operational, and tactical capacity of each nation. This delineation is useful for two reasons. First, it provides a way to look at four key areas of military endeavor that are each distinct but also are closely linked in the overall approach to a nation's military activities. This ensures that action at each level supports overall strategy and achieves desired national objectives.

A second reason why this is a useful analytical framework is because it neatly aligns with the doctrine of most military institutions. Military doctrine makes a very clear delineation between tactical, operational, and strategic activities. This can assist senior military leaders in visualizing the sequencing and prioritization of military activities and provide a framework for ensuring that military actions meet desired strategic and political objectives. But what are these levels, and how are they related?

Starting from the foundations of warfare, the tactical level is where battles and other forms of engagements between opposing military forces are conceived, planned, and executed in order to achieve military objectives. Tactical forces employ tactics to achieve the outcomes they have been given, generally in the form of a mission statement from a superior headquarters.

Next is the operational level, at which tactical objectives and actions are sequenced and orchestrated—often as campaigns—to achieve military and strategic objectives. Importantly, much of the prioritization for allocation of forces, logistic support, intelligence, transport, and interdomain collaboration takes place at this level. The operational level is also largely joint rather than service-oriented. It acts as the interface between the tactical and strategic levels of warfare.[25]

The next level is the strategic level of war. While strategy in the 1800s and early 1900s was almost an entirely military affair, that is no longer the case. After World War I, strategy became a national, rather than a military, endeavor. Many military institutions also subcategorize strategy. In U.S. military doctrine, the strategic level is split into national policy (civilian-led) and theater strategy (military-led).[26] Other countries such as Australia have split it into the national strategic (political dimension and civilian-led) and military strategic (military planning and direction).[27] In the United Kingdom, the subcategories consist of national strategy, defense strategy, and military strategy.[28]

While it is possible to distinguish between these three layers in a doctrinal publication, in practice they overlap and interact. Each layer has an influence on the others. Strategic requirements will influence operational goals and tactical actions. Concurrently, tactical success or failure can also have an impact, sometimes in disproportionate ways, on strategy. The potential for the overlap and interaction between these different layers will also be considered as we define the notions of military effectiveness in each level.

A hierarchical approach to military effectiveness is useful. It can be informed by themes that pervade all of the levels. For example, we have already identified seven trends in future warfare that can be applied in developing the framework for military effectiveness. The four themes of integration, responsiveness, skill, and, quality provide additional depth to the analytical framework to ensure military institutions' effectiveness in the twenty-first century.

However, one final aspect must be considered in this exploration: the pace of change in the environment. A consistent theme throughout this book has been the surge in innovation and the speed of change in the strategic environment. This means that military organizations must not only be effective at each level, but they must also be effective at adapting more rapidly to change—in peace and in war. Consequently, in addition to military effectiveness at the strategic, operational, and tactical levels, the ability to adapt at all levels will have a significant impact on military effectiveness and competitiveness. To that end, the capacity to adapt must be examined as a fourth and separate element of military effectiveness.

## STRATEGIC EFFECTIVENESS

War remains an enduring aspect of human existence. National governments must assume that they will need to defend their territory and their sovereignty. To do so, they must invest in military institutions. Any notion of effectiveness for these military organizations must be drawn from political purpose. There must be an alignment of what political outcomes are sought and what military strategies are developed, executed, and adapted (primarily a military activity). Military strategy is an integral part of higher-level national strategy, designed to meet the policy needs of government.

Strategy is a word, and a concept, that has resisted a single, agreed definition. As Beatrice Heuser writes, "Strategy is hard to press into one universally accepted definition accepted throughout the ages."[29] Hew Strachan has described strategy as a word "used by governments to describe peacetime policies more than by armies to shape wars" and that has "gained in breadth but has forfeited conceptual clarity."[30] Colin Gray describes strategy as "a bridge between purpose and action."[31]

Strategy comes into play when there is potential or actual conflict, where the interests of two or more actors collide, and where some type of resolution is needed.[32] A central idea in the theory and practice of strategy, and therefore in achieving strategic effectiveness, is that it exists in an environment where actors are competing and where there is some misalignment of larger objectives. As Heuser notes, "Strategy is a comprehensive way to pursue political ends, including the threat or use of force, in a dialectic of wills—there have to be at least two sides to a conflict."[33] This capacity for strategic thinking is especially compelling given that the complex problem of running military activities is liable to occupy the skills and minds of senior commanders so completely that it is easy to forget what it is being run for.[34]

Millett and Murray have described the capacity for effective strategic thinking as being much more important than tactical or operational excellence.

They stress the importance of getting strategy right (and the strategic education that enables this) when they state that "it is more important to make correct decisions at the political and strategic level than it is at the operational or tactical level. Mistakes in operations and tactics can be corrected, but strategic mistakes live forever."[35] For this reason, strategic effectiveness is of profound importance to twenty-first-century military institutions and the nations to which they belong. Of all the levels of effectiveness explored in the following pages, this is perhaps the most vital.

### Integration

A more integrated approach to national security will be required in the decades ahead. Brooks and Stanley reinforce this point, describing "integration" as one of their core elements of military effectiveness.[36] Western military organizations must continue to develop the mechanisms by which they link purpose and action. Military institutions must be able to achieve desired military strategic objectives that complement economic, information, cultural, diplomatic, and other strategic goals. These must be combined to support a nation to secure its political goals.

The alignment of military strategy with other elements of national security demands that military leaders be politically aware (but not political). They must be able to sustain an effective and continuous dialogue with government. This dialogue, while central to effective civil-military relations, is not just about sustaining relations in accordance with the laws and norms of given countries; there also are important outcomes from it for strategic military leaders. They must be able to generate sufficient influence with political leaders so that they seek logical military goals, which must be consistent with the capabilities, size, and posture of military forces and be supportable by the logistic and industrial capacity of the nation. At the same time, strategic military objectives must possess a level of alignment with security partners and allies.[37]

The British in 2020 described something like this in their *Integrated Operating Concept 2025*.[38] To realize such an integrated approach, military institutions require the organizational mechanisms and procedures that can achieve this closer integration of national power. Therefore, the first element of strategic effectiveness for twenty-first-century military institutions is the alignment of national policy and strategy with military strategy in order to achieve desired defense and deterrence outcomes.

Without a pragmatic and integrated approach to policy and strategy, it is difficult to imagine how a nation might fare in addressing the challenges of the twenty-first century. But even if this aspiration is achieved, it is insufficient to assure national sovereignty in the decades ahead. The successful execution of national policy and military strategy will require all the institutions of national security to develop their relationship with, and exploitation of, information.

### Information

Twenty-first-century technologies are providing methods for influencing different populations in a way that has not been possible before. As such, military organizations must be able to leverage information to keep their own government and societies informed. Transparency and auditability are core responsibilities of military institutions in democratic societies. Military organizations also need the capacity to influence competitors and adversaries while countering the influence activities of those same entities.

The use of propaganda and disinformation by military organizations has long historical precedents. What is new, however, is the ease, global reach, speed, and low cost of such activities. The barriers for entry into the conduct of information activities are low. As such, they have become a core tool for both state and nonstate actors.

In their 2018 examination of social media and information operations, Peter Singer and Emerson Brooking propose several "rules" for this evolved information environment. One of them is that the Internet is now a battlefield, which changes how we must think about the nature of information. Further, because of how the information environment has evolved, war and politics have never been more intertwined.[39] It is an elegant summary of the challenge faced by societies, governments, and military institutions in the contemporary environment.

Therefore, another measure of the strategic effectiveness of twenty-first century military institutions will be their capacity to undertake information operations and generate what the British call an information advantage.[40] Successful military institutions in the coming decades must be able to leverage information to inform or enable their operations and deny information to adversaries while also building resilience in their institution. These resilience activities must be nested within, and aligned with, national approaches to resisting foreign disinformation and influence campaigns.

### Resources

Military institutions need resources. Indeed, they are extraordinarily large resource-consuming organizations that demand a significant share of national wealth to build and sustain their capabilities. During war, the need to fund personnel and technological resources exponentially increases this strain on the national coffers.

Leaders of military institutions must be able to make the case to their political leaders for sufficient budgets to build the military capacity that meets the policy objectives of their governments and the national security requirements of their nation. When there is no obvious threat, this is difficult. The 1990s and the interwar period serve as examples of where military institutions sometimes failed to offer compelling and evidence-based cases for sufficient investment. However, where there is a clear threat to a nation's interests or sovereignty, the case for sufficient resourcing is clearer.

Military institutions also need people. Their leaders must be able to justify an appropriate overall force size to their national leaders. Sub-elements of this will be the balance between different services and between regular and reserve forces. Providing the right evidence for force levels desired by the strategic leadership of military institutions involves both the art and science of our profession; long-term workforce planning, careful experimentation with existing and new workforce specialties, recruiting strategies, and sustaining an appropriate level of retention are all components.

The final resource is access to the technological (and industrial) capacity required to produce equipment and other elements of military capability. While technology rarely provides a silver bullet for military success, military institutions must be able to secure access to the right amount of high-technology weapons, communications, and other equipment to enable their activities. As Millett and Murray note, "A military organization that cannot or does not exploit either domestic or foreign industrial and scientific communities limits its effectiveness."[41] Therefore, effective twenty-first-century military organizations must seek and gain access to the technological and industrial capacity that underpins their capabilities. This requires investment in research and development, partnerships with civilian universities, collaboration with domestic and foreign industrial partners, and the sharing of sensitive technologies with close allies and partners.

In wrapping up this discussion on resources, the third element of strategic effectiveness for twenty-first-century military institutions is their ability to secure for themselves regular and sufficient allocations of a nation's budget, people, and technological and industrial capacity to build competitive military forces and capabilities.

### Intellectual Competition

Military institutions, working within a more integrated national security environment, must adapt to new technologies and other capabilities being fielded by our state-based competitors as well as those that might be wielded by nonstate actors. Besides the new high-velocity and low-signature weapons systems—potentially fielded in large quantities—an array of nonkinetic capabilities such as cyber, quantum encryption, stealth technologies, and influence activities must be dealt with.

Accompanying these new systems are new ideas about how to use them. Both the Chinese and Russians have invested in new operational concepts that are designed to attack Western systems and joint forces where they are weak. Therefore, the fourth element of strategic effectiveness for twenty-first-century military institutions is investment in winning the intellectual competition with strategic competitors and adversaries. Military institutions must be capable of generating competitive strategies that pit their strengths against potential adversary weaknesses, align with national strategies, and complement those of allies and security partners.

### Assessment

Winning the intellectual contest in future competition and conflict will rely on evolved ways of analyzing the strategic environment. The United States has a multi-decade history of undertaking net assessments of its principal competitors and using the knowledge gleaned from these to build competitive strategies. More recently, the United Kingdom established its Strategic Net Assessment capability within the Ministry of Defence in 2018.[42] All Western nations need a capacity to undertake these kinds of assessments, and these assessments must be undertaken from national as well as alliance perspectives.

This approach is a well-worn path in the U.S. Department of Defense's Office of Net Assessment. As Paul Bracken has written, "Net assessment had its origins in the need to integrate Red and Blue strategy in a single place."[43] This was driven by a growing dissatisfaction with existing analytical tools within the U.S. military establishment in the late 1960s. In the wake of a 1971 study on U.S. intelligence operations by National Security Council staff, a recommendation was put forward to establish a "net assessment staff." That November, the president signed a directive to establish this new activity, and Andrew Marshall formed the new net assessment group in the spring of 1972.[44] Developed throughout the early 1970s and 1980s, the group offered a way of thinking and a process for comparing the United States with its principal adversary that evolved over the following decades.

What might these strategic net assessments look like in the contemporary era? One of the most important elements that strategic assessment must explore is the nature of the competition between nations. The appreciation of why states compete provides insights into areas where they might take action to either reduce tension or ameliorate the impacts of that competition. This description of the competition should include how both sides see the different areas of competition as well as the importance attached to each.[45] It also includes political systems, objectives, strategic cultures, differences in strategic competencies and characteristics, and the impact of interaction between competing nations.

Another element of these strategic assessments is the development of hypotheses about the future security environment in which military forces will operate. There are a range of different futures methodologies currently used by military organizations in countries such as the United Kingdom, the United States, Singapore, Australia, and others. Products include the British Defence Concepts and Doctrine Centre *Global Strategic Trends* and the U.S.

National Intelligence Council *Global Trends* reports.[46] They provide trends analysis, alternative scenarios (an approach pioneered by Shell in the 1970s), and detailed analysis on the potential impacts on military operations and institutions. This is vital in developing future force structures and concepts and in exploiting new technologies as well as developing competitive strategies at the national level.

Therefore, the fifth element of strategically effective military institutions in the twenty-first century will be their ability to undertake rigorous strategic assessments, including hypotheses about the future security environment, upon which they might base the development of competitive strategies. These assessments and competitive strategies will be significantly more effective if they are prepared, executed, and adapted in partnership with other like-minded nations. As such, strategic engagement comprises another measure of strategic effectiveness.

### Engagement

Engagement between like-minded military institutions and like-minded nations must continue to evolve and embrace a greater sharing of ideas on every aspect of military endeavors. This will include strategy and joint operations but will also extend to logistic support, technological research and development, and the full spectrum of personnel training, education, and development. Strategic engagement between Western military institutions already includes alliances such as the North Atlantic Treaty Organization (NATO), the Australia–New Zealand–United States Treaty Alliance, the U.S.-Japan Treaty of Mutual Cooperation and Security, and the "Five Eyes" arrangement between the United States, Britain, Australia, Canada, and New Zealand. But engagement across a wider range of activities and with more nations in the Indo-Pacific and beyond will be required in the coming decades.

Strategic engagement beyond military institutions is also vital. Industry-military links have long been critical in the development of military capability and are generally robust in Western nations. These relationships remain important for the provision of cutting-edge equipment and other services. However, this only provides physical capacity. Military institutions can better augment their intellectual capacity through strategic engagement activities.

Therefore, the ability to undertake enhanced collaboration with civilian universities and think tanks is also essential. These institutions provide a level of diversity in intellectual capacity that is almost unthinkable in military institutions. In essence, engagement between the military and academia should ensure a broader and more diverse range of strategy options to be considered by military leaders. While there are a range of collaborations with universities on the technological front, collaboration on new strategies and ideas could be significantly improved. It demands more investment across Western nations and represents a low-cost, high-return activity compared to investments in advanced technologies. Consequently, the sixth element of strategic effectiveness for twenty-first-century military institutions is the capacity to strategically engage with civilian industry and academia in a more seamless and collaborative way, to produce new ideas and technology that will provide a competitive military edge.

### Sustainability

Aligning strategy and policy, securing resources, investing in intellectual competition, and strategic engagement are vital. However, even in their optimal combination, they remain insufficient to build effective twenty-first-century military organizations. Another element of strategic effectiveness involves the long-term sustainability of military institutions in future warfare. This has several components.

First is a nation's ability to logistically support its military forces through sovereign industrial output. The capacity of nations to provide critical supplies, particularly medicines and protective equipment, came to the fore during the COVID-19 pandemic. Effective military institutions must assure themselves of enduring access to critical military supplies during peace and war. This requires arrangements with national industries and with allies. The coercive behavior of China in 2020 is also driving nations to reexamine their internal capacity for manufacturing a range of products, many of which have dual civilian and military uses (including food, fuel, chemicals, computing technology, AI, biotechnology, and pharmaceuticals).

A second element of military sustainability is the capacity of a nation to mobilize an expanded military in a reasonable amount of time—that is, before that nation is defeated by an adversary. The subject of mobilization has received greater attention in many Western military institutions over the last several years—and so it should. Military institutions are expensive in peacetime. Wars, however, impose much larger human and financial costs and are decided as much by attrition and exhaustion as they are by the will of the belligerents. The planning for this must start well in advance of any hostilities.

A 2020 study conducted by the Australian National University has explored mobilization issues from an Australian perspective. Three separate papers approach the topic from different perspectives, but clear themes emerge. First, mobilization is a national (and potentially an alliance) issue, not a military undertaking. Nations need to decide which national and military capabilities and supplies will be developed indigenously, which ones will be sourced on a shared basis with friends and allies, and which ones might be procured off the open market. Second, it demands a rational assessment of what is possible through mobilizing industrial capacity for military ends. Most likely, it will be scaled against the size of the external threat; total mobilization is unlikely except for the most perilous of threats to a nation. Finally, mobilization is a social issue—taking hundreds of thousands (if not millions) of people from civilian to military occupations is a statement of national will.[47]

While industrial output has been a long-standing subject of study in strategy, the events of the past two years have brought to the fore the importance of national sovereignty in some forms of critical manufacturing capabilities. In the wake of COVID-19, a range of countries are reassessing the costs and benefits of importing goods versus manufacturing them domestically. Developing domestic sources of supply, especially for technologically advanced systems, is both time-consuming and expensive. But if nations wish to possess greater surety in supply in certain classes of manufactured items, the concept of national resilience will be an important consideration in the development of national security strategies.

In his 1973 book *The Causes of War*, Geoffrey Blaisey writes that "it is doubtful that any war since 1700 was begun with the belief by both sides that it would be a long war. No wars are unintended or accidental. What is often unintended is the length and bloodiness of the war."[48] Wars nearly always last longer than expected and are generally more expensive and exhausting than the belligerents are prepared for. Therefore, the next element of strategic effectiveness for twenty-first-century military institutions is to effectively plan and design systems for strategic resilience, industrial sovereignty, and mobilization.

Future victory or possibly national survival will hinge on how well a nation can harness both its material and its moral strengths. A significant contributor to how it might do so is the focus of the final element of strategic effectiveness: organizational culture.

## *Culture*

Culture has a significant impact on military institutions. It influences military organizations' success and failure in all their activities. Cultural factors determine the professionalism and discipline of individuals and teams in military institutions. These factors drive battlefield and broader military effectiveness.

The rise of the power of Western nations was due in large part to changes in military culture since the seventeenth century. Before this time, military institutions were often masses of ill-disciplined individuals and groups. Beginning with the seventeenth-century reforms of Maurice of Orange and Gustavus Adolphus, military institutions reformed and imposed strict professional discipline, which in turn assisted in the creation of the modern state. As Murray notes, "It is military culture, rather than technology, that explains the extraordinary record that Western military institutions have achieved over this period."[49]

In their landmark study of military culture, *The Culture of Military Organizations*, Peter Mansoor and Williamson Murray have defined organizational culture in this context as "ideas, assumptions, norms, beliefs, rituals, symbols, and practices that determine how the institution functions and give meaning to its members."[50] Importantly, culture in military institutions establishes distinct organizational identity (and sub-identities) and expectations about how members of the institution will act in given circumstances. It is shaped by external factors such as geography, history, and the strategic culture of the nation to which the given military institutions belong.

The culture of a military institution is foundational to its overall effectiveness. Almost every action taken by individuals and groups in a military organization is in some form shaped by its overall military culture.[51] There are five elements of culture that will underpin successful military institutions in the decades ahead. The first is that all members of a military institution must be imbued with a service ethic. Their first duty is service to their nation. Part of that service is adherence to the values of that nation, including its laws and norms around ethical conduct and the use of force. A second element of an effective military culture is a focus on professional excellence, at the individual and collective level. This is honed through training, education, experience, and good leadership. A third desirable element is that institutions must be capable of honest studies of military history and future challenges that then can be exploited to develop the concepts, structures, and equipment of the institution. This requires building diverse viewpoints within the organization, as well as investment in education and the organizations that can analyze operational and tactical lessons. It also demands sufficient bureaucracy to ensure the maintenance of key functions, but not so much that it gets in the way of innovation and rapid adaptation where required.

There are two final components of an effective military culture in the twenty-first century. The first is that institutions must be learning organizations. From top to bottom, the incentive systems (promotion, medals, and so forth) of military organizations must nurture creativity, self-critique, and new ideas, which mid-level leaders can invest in and senior leaders can champion. This culture must nurture individuals who can creatively outthink and outplan potential adversaries. At the same time, there must be a cultural affinity with harnessing the disparate and diverse intellects of all its people to solve complex institutional problems in the short, medium, and long term. This must be applied to force design challenges, operational concepts, the integration of kinetic and nonkinetic activities, and personnel development and talent management. Frank Hoffman has recently made a detailed exploration of the ability to leverage learning by military institutions. In *Mars Adapting*, he describes the process as *organizational learning capacity*.[52] This capacity is a vital part of the culture of military institutions.

The final component of an effective military culture in the twenty-first century is excellent leadership. Leadership is the art of influencing and directing people to achieve organizational goals. The best military organizations emphasize this as a central element of the military profession where leaders trust and are trusted, nurture innovation, build diverse and capable teams, learn from many and varied experiences, and strengthen their understanding of ethical decision-making. Good leadership is a defining element of successful institutions. The outcome of good leadership is a military institution that is not only effective but also legitimate in the eyes of the government and the civilian population.

Rapid change in the environment is also driving developments in command and control that embrace greater decentralization of decision, self-synchronization, and self-organization. This allows an integrated military organization to more rapidly reorient toward new and evolving missions and is essential at every level of military institutions.

While military institutions around the world have many common traits—uniforms, the use of force, ranks, traditions, hierarchies, and such—they are also uniquely influenced by their national cultures, conventions, history, and geography. There is a careful balancing act between national and military cultures. In a democratic system, we must never allow military cultures to override those of the nation they serve. This is the pathway to civil-military problems. Therefore, effective military institutions in the twenty-first century must develop and nurture an organizational culture that is subordinate to national culture, encourages professional excellence, possesses a strong learning culture, and incentivizes ethical leadership that is legitimate in the eyes of government and the people.

## OPERATIONAL EFFECTIVENESS

The operational level of warfare links tactical actions to strategic outcomes. The theory of the operational level of war and operational art has been the topic of significant investigation and debate over the last several decades. While a small minority retain some skepticism about this as a level of war, Western military institutions have reached a consensus that it does exist, and they possess doctrine in this element of the military art.[53] It is therefore appropriate that effectiveness at this level of war be part of our analytical framework for effective twenty-first-century military institutions.

Because a common definition of the operational level has been broadly agreed among NATO allies, it is perhaps the most relevant one for our purposes here. NATO doctrine defines the operational level as "the level at which campaigns and major operations are planned, conducted and sustained to achieve strategic objectives within theaters or areas of operations."[54] Tactical operations by joint forces involving position, logistics, intelligence, influence operations, maneuver, and other tactics are prioritized, orchestrated, and combined so that desired objectives can be achieved at the strategic level. A large proportion of planning and command at the operational level is highly mechanistic, requiring staff coordination and synchronization. However, there is another aspect to the operational level, which is known as the operational art.

Operational art is the creative and skillful use of tactical means in order to achieve strategic ends. This demands the resolution of the tension between tactical realities and strategic demands through a continuous process of design, planning, and execution of operations.[55] But the fact remains that the most important function of the operational level is to ensure that tactical actions are orchestrated in place, time, and resources to achieve strategic outcomes. If this is not the case, the chosen method of operational design is inappropriate or potentially negligent. Therefore, the first element of operational effectiveness for twenty-first-century military organizations is that their operational concepts must align with the strategic objectives that have been assigned to them.[56]

Operational art provides an overarching framework for the operational level of warfare to align with strategic objectives. It also provides the intellectual power and imagination to outthink an adversary, deny them their desired goals, and destroy or interfere with their operational design and plans. It is this intellectual battle that drives the second element of operational effectiveness for twenty-first-century military institutions: operational concepts.

### Operational Concepts

Nearly twenty years ago, Hans Binnendijk and Richard Kugler produced an insightful paper exploring the pathways for the U.S. military to transform itself for the challenges of the twenty-first century. At the heart of their argument was the need for new and evolved operational concepts: "If defense transformation remains anchored in old concepts, it risks perpetuating the status quo, even if it alters forces and weapons."[57]

The requirement for effective operational concepts was also a central aspect in Millett and Murray's three-volume assessment of operational effectiveness. Therefore, a second element of operational effectiveness for military institutions is their capacity to generate the ideas and concepts that link strategy and tactics, to test them thoroughly, and to implement them in concert with new technologies, organizations, training, and education.

An important requirement of operational concepts is that they must pit the strength of Western military forces against the weaknesses of their competitors and potential adversaries. A recent historical example of this would be the United States targeting Iraqi operational command and control (an assessed weakness) with its overwhelming dominance of airpower (a U.S. strength) during the 1991 and 2003 Gulf wars. In the contemporary environment, the PLA has assessed that a key weakness in Western military organizations is the operating systems that link forces in the different domains and their supporting logistics, intelligence, space, and information systems. Based on their close observation of joint military operations since the 1991 Gulf War, military theorists in the PLA developed the theory of operational success called systems destruction, which we explored in the previous chapter.

The concept of systems destruction relies on the possession of a range of operational systems that are sufficiently multifunctional and multidimensional to confront Western military systems in all domains concurrently. It seeks to build a system comprised of *elements* (command and control, firepower, intelligence, information, etc.), *structures* (a matrixed work organization with all systems linked through information technology), and *entities* (the smallest units within the operational system). Applying these in a highly integrated network, systems destruction warfare then aims to paralyze an adversary's operational system.[58]

Western military organizations must match and improve on this Chinese conceptual development. Testing new military concepts must include open debate within and outside military institutions. A historical example of this is the multiyear institutional debate in the U.S. Marine Corps in the 1980s that underpinned its transition from *methodical battle* to *maneuver warfare*.

In the contemporary environment, the profusion of blogs, seminars, and other information technologies of the past two decades must be exploited to enable this debate. Experimentation, the creation of virtual avatars of existing military forces, classified wargames, and live exercises also must be used to identify the strengths and weaknesses of new operational concepts, force structures, service balances, and interdomain cooperation.[59] Testing is time-consuming and resource-intensive, but it is nowhere near as expensive as going to war, or being in a strategic competition, with the wrong operational concepts. The process of testing can also expose large parts of military institutions to new ideas and underpin buy-in from across the force for new operating concepts.

These new operational concepts must also be integrated with other concepts as part of a family of operational ideas. Like weapons, there will never be a single operational concept that will cover the breadth of military activities at the operational level. Also, like weapons systems, joint operational concepts are most effective when they are integrated with other operational concepts. The aggregation of operational concepts should overwhelm a potential adversary's capacity to think, plan, and act at the operational level and deny them the ability to attack our systems. Integration of this sort is a multidomain undertaking, necessitating land, sea, air, space, and cyberspace integration. It is also a functional integration issue, incorporating military capabilities such as air defense, logistics, intelligence, command and control, and influence operations. Finally, new operational concepts must be integrated across different nations to ensure mutual support and understanding, complementarity across slightly different approaches and systems, and the capacity to generate unified action at the operational level.

Since the end of World War II, various nations have invested in their development of the theory and practice of joint operations. Military institutions must now take another step forward in developing the next generation of joint operations and joint operational concepts. They must cover the operation of military and other elements of national power in all domains concurrently. New concepts must be tested, integrated, supportable, and aligned with available technology. Therefore, another measure of operational effectiveness is how well new operational concepts across the spectrum of strike, mobility, targeting, deep operations, missile defense, and other areas are developed and then rigorously tested and exercised in the field regularly.

### Matching Concepts to Capacity

Operational concepts are important, but there is little point in developing ones that are beyond the military capacity of a nation. Consequently, part of the testing of operational concepts must be their validation against the intelligence, logistics, personnel, transport, infrastructure, army, air force, navy, and other capabilities possessed by the military institution.[60]

The technological capabilities of a particular military force must be adequate to achieve the objectives of new operational concepts. Where new or disruptive technologies appear, they must be considered by new operational concepts. Artificial intelligence is likely to result in different ways of conveying information and sharing lessons. It will probably support decision-making at all levels of military endeavor. It will also significantly affect cyber operations and the conduct of influence operations.

Hypersonic weapons are likely to change the pace of operations at the operational level, requiring new and more rapid ways of communication, analysis, and decision-making. Autonomous systems in the air, on the land, and at sea will change how logistic support is undertaken and how joint forces are structured to move, deny, or seize territory through fires and presence, to fight, and to influence the activities of an adversary. All these are foreseeable now. It demands that military institutions refocus their energies on new ways of thinking if they are to be effective operational forces in the twenty-first century. Therefore, new operational concepts must also be supportable by the means available to the military institutions of individual nations and align with available technology.

### *Flexibility*

Finally, for new operational concepts to be most effective, they must also be flexible enough to adapt to the changing character of war. The interaction of two military forces, during strategic competition or in war, drives rapid adaptation in tactics, concepts, equipment, training, and leadership. Consequently, the third element of operational effectiveness for twenty-first-century military institutions is their ability to anticipate surprise or change and adapt their operational concepts to retain their effectiveness for current and future operations. This will be explored in more detail later in this chapter.

## TACTICAL EFFECTIVENESS

As a young staff cadet at the Royal Military College in Canberra, I spent days in the military theater receiving lectures on tactics. We spent many more days out at the Majura Training Area, just outside of Canberra, undertaking tactical exercises without troops (TEWTs). These provided my classmates and me with on-the-ground experience in the application of what we had learned in the lecture theater.

Leadership and tactical acumen were central to being an effective junior officer in the Australian Army. Professional courses at our Land Warfare Centre in Canungra and more TEWTs at the company and battalion levels served to hone my tactical knowledge. It is a foundational skill for military leaders in any service and in any nation.

At the most basic level of military operations, armies, navies, air forces, and their supporting elements must be able to fight and win battles. While this may now incorporate space-based capabilities as well as nonkinetic actions such as cyber and influence activities, the fact remains that military forces must be raised, trained, and sustained in peacetime to form a credible deterrent. In war, they must be able to fight and win battles and campaigns—period. The tactical level is focused on the planning and employment of military forces in battles, engagements, and other activities to achieve military objectives. Normally, these tactical actions are undertaken within joint task forces, but in rare circumstances, military activities by a single service might be undertaken. Important foundations for tactical action are tactics and leadership.

Tactics have both artistic and scientific elements. The artistic realm of tactics lies in military leaders' capacity to apply imagination to the use of the means available to them—forces, weapons, and procedures—in order to seize, retain, and exploit the initiative against an adversary. Another element of the art of tactics is the construction of cohesive, combined arms, joint units that are able to successfully execute tactical activities. This demands a deep understanding of the moral, ethical, and physical impacts of tactical actions on military personnel. Fear, friction, uncertainty, and Clausewitz's "fog of war" are most proximate at this level of war. It demands well-honed leading and planning capabilities in all military leaders who are able to sustain tactical cohesion under the worst of circumstances.

The science of tactics incorporates a deep understanding of the capabilities and effects of one's own weapons as well as those of the enemy. It also includes understanding the different tactical formations and specialties that comprise combined arms, task force, and joint organizations and how these can be swiftly reorganized and retasked depending on rapidly changing tactical situations. Finally, the scientific element of tactics incorporates an appreciation, and constant rehearsing, of the procedures and techniques required to achieve specific missions. These might include drills for obstacle crossings, quick attacks, relief in place, defensive activities (in all domains), and other tactical actions. The artistic and scientific elements of tactics have been an essential aspect of learning for military leaders throughout history. This will remain the case in the twenty-first century.

A final aspect of the tactical art is leadership. Good leaders are good tacticians who also possess the presence of mind to remain calm in the most austere and difficult circumstances and can make good decisions about their people in uncertain, rapidly changing environments.[61] Those who can keep their head in the awful chaos of battle are more likely to win than those who cannot. However, military leaders need to master the scientific aspects of tactics as well.

### *Dispersed but Unified Forces*

The history of tactics in the past century has largely been a story of the increased dispersal of forces. This trend has been driven primarily by the enhanced capacity of military forces to find their enemy and to attack them with a greater weight of fire and with significantly improved precision. While this trend was under way in the 1860s and 1870s, it matured during the world wars of the twentieth century and reached its ultimate expression with reconnaissance and strike complexes guided by GPS during two wars in Iraq and Kuwait in 1991 and 2003.

The continued improvements in reconnaissance and strike capabilities have forced military institutions to spread their forces and people over greater distances. The German solution to this problem in 1917–18—mission command and mission tactics—was to emphasize decentralized command and control, allowing junior commanders down to the lowest level to make tactical decisions on their own and in accordance with higher intent rather than tasks.[62]

Throughout history, forces that have debuted new tactics and doctrine have often had the advantage of surprise. As Trevor Dupuy notes in his book *The Evolution of Weapons and Warfare*, "Twice within the lifetime of men now living, the German Army has scored stunning tactical surprises over its opponents—in 1918 and in 1940—yet in neither case did it use new weapons. Every item in the German arsenal was familiar, yet revolutionary use of these weapons came as a great surprise."[63] New tactics and doctrine therefore provide a much greater return on investment for military institutions than new weapons systems.

The trend in dispersed tactical forces continues into the twenty-first century. Precision weapons and dense networks of advanced sensors result in a much more lethal environment. This means that the advantage in the contemporary environment has shifted to defensive tactics. As one of the foremost contemporary experts on this topic, Robert Scales, notes, "A battlefield dominated by firepower and the defensive compels units to disperse, disaggregate, and go to ground. Disaggregation is good in that it lessens the killing effects of firepower but bad because dispersed forces are less able to mass, and mass is essential if maneuver is to be restored. The purpose of small units changes on a distributed battlefield. In the future, small units will become virtual outposts, in effect the eyes and probing fingers of a larger supporting operational force placed out of reach of the enemy's long-range fires."[64]

Some contemporary military institutions have experimented with this more distributed form of tactical operations. The U.S. Marine Corps has undertaken multiple experiments to examine different unit structures and capabilities that might be required to achieve a disaggregated effect in future operations. Its 2005 *Concept for Distributed Operations* is described as a concept to "create advantage through the deliberate use of separation and coordinated, interdependent, tactical actions."[65] The arrival of Gen. David Berger as commandant in 2019 heralded even greater transformation. His strategy to redesign the Marine Corps to provide for small, hard-hitting, and distributed forces envisions a fundamental shift in the design of the Marines. It features the disposal of larger, harder-to-deploy, and higher-signature platforms such as tanks and towed artillery and increased investment in long-range rocket artillery and unmanned platforms.[66]

The Russians have also been active in this area. Their operations in Ukraine in particular have highlighted doctrinal innovation in the structure and operations of their tactical forces. These tactical changes align with an evolution in strategic thinking driven by Gen. Valery Gerasimov called new-generation warfare. Encompassing almost the entire breadth of activities in warfare (less weapons of mass destruction), it includes cyber, influence, private military forces, and different tactical units. The most prominent new tactical force is the battalion tactical group (BTG).

First employed in operations in Donbas, BTGs were formed by almost every Russian field army and corps and deployed for operations around Ukraine. Comprised of tanks, mechanized infantry, air defense, artillery, and anti-armor troops, these battalion-sized units possess more firepower than most Western brigade-sized formations. The BTGs have proven to be highly resilient and quite deadly, able to bring massive amounts of firepower to bear on an adversary in very short order. This new form of disaggregated tactical operations has proven so successful that the Russian army plans to double the number of BTGs from 66 to 120.[67]

These changes are not without some risk. Dispersing forces means that military institutions must trade efficiency for survivability. Large, common organizations that are concentrated are easier to command, provide for better mutual support between units, and are simpler to support logistically. But they also have a large targetable signature, which must be reduced if survivability of the force is to be enhanced. Additionally, without robust, secure communications networks, new ways of resupplying distributed units, and the provision of well-trained leaders able to apply mission command, these new forms of tactical units and maneuver are unlikely to be successful.

In the decades ahead, concentrated bodies of troops, regardless of leadership or tactical acumen, will have easily detectable signatures and will be very attractive targets for enemy recon-fires complexes. Therefore, the capacity to operate in a unified fashion but as much more dispersed tactical forces to defeat signature detection and enemy lethality will be an important measure of twenty-first-century tactical effectiveness.

## Mission-Focused Organizations

The conceptual work being undertaken by the U.S. Marines and the Russians emphasizes not only small signature tactical organizations but also the integration of different capabilities under a unified command structure. The combination of the different "arms and services" within a military institution, usually called combined arms, was developed in its modern form on the Western front by allied forces to smash through German defenses in 1918. The combination of armor, artillery, infantry, engineers, communications, logistics, and aircraft featured heavily in the 1918 August offensives. In the interwar period, it was elevated to an art form by the Germans and used most effectively in France in May and June 1940.

At the same time, the United States and the United Kingdom were developing concepts for the integration of land, air, and sea power (amphibious operations). As the war progressed, combined arms and air-land integration became more sophisticated. The latter half of the twentieth century saw this integration elevated to the operational level, with joint task forces becoming more capable and more frequently used in operations in the Middle East and beyond.

In the twenty-first century, space, cyber, and influence operations are essential elements of any form of joint or coalition activity. These functions must be integrated better into existing forms of tactical activities. Therefore, the capacity to build mission-focused organizations, which combine a variety of air, land, sea, cyber, and information capabilities and then wield these effectively, will be another measure for effective tactical organizations in the twenty-first century.

### *Aligning Activities with Objectives*

More dispersed, combined arms joint forces must conduct their activities so that they achieve desired outcomes of operational commanders and strategic leaders. Strategic and operational objectives provide vital context for planning, executing, and measuring the success of tactical actions by military forces. They must guide the undertaking of operations and, thus, of tactical activities.

This comes with a caveat, however. Strategic aims and operational goals must be in alignment with the tactical competence and capabilities of military forces. For example, a strategic objective that insists on achieving a military strategic deterrent effect would be ineffective if that nation's military institutions lack any form of offensive long-range strike, intelligence collection, or strategic influence capabilities. As Williamson Murray writes, "What is tactically feasible should shape the selection of strategic objectives and plans."[68]

In his 2017 book *On Tactics*, B. A. Friedman explains many of the most vital tactical approaches and places them in a framework of physical, moral, and mental tenets.[69] In the book, he connects tactics to policy through the use of strategy: "Overwhelming military success was once enough to bridge the gap between tactics and policy, and thus was a sufficient strategy. Military success alone is no longer enough. . . . Military success is only the means to an end. If we do not understand our end, no means will accomplish it, and the battle becomes only slaughter."[70] To that end, a third measure of tactical effectiveness for twenty-first-century military forces will be their ability to carefully align tactical activities with operational and strategic objectives.

## *Training and Education*

Another area of tactical focus is the development of people. Effective tactical organizations, even with the best equipment and the most rigorously thought-out techniques and procedures, are useless without well-trained and -educated people at every level. The last two hundred years of warfare—whether large-scale massed conventional forces, low-level insurgent campaigns, or influence campaigns—have demonstrated that highly developed training systems and a focus on good leadership founded on mission orders provide a better chance of success.

Well-trained personnel can perform their duties at a high level in the arduous and confusing crucible of combat. Instinct, born of years of individual and collective training, provides the essential tool for military personnel to endure and overcome the fog and friction of war. At the same time, well-selected, -trained, and -educated leaders—officers and noncommissioned offers—provide the discipline, tasks, unit cohesion, and purpose that are essential in effective tactical forces. Institutional training systems cannot be static, however. They must constantly evolve to take account of changes in weapons systems, enemy threats, new tactics, and changes in the broader strategic environment.

The rapidly evolving capabilities of artificial intelligence, coupled with new robotic systems, hold the promise of achieving more effective outcomes from military operations, underpinned by better decision-making by our military leaders and their political masters. Humans must now be able to work in concert with swarms of robots as partners. They will often do so in environments and organizations where there may well be five, ten, twenty, or even hundreds of robotic systems for every human being.

Contemporary military institutions possess limited doctrine for this emerging technology. They also lack fully developed training and education regimes to ensure that both the people and the robotic systems in this new environment are optimized for mission success. Those nation-states that are early adopters and have the vision to embrace this evolving technology while in its embryonic state will win themselves a significant military advantage into the future.

Moreover, the partnership between humans and algorithms (represented by what we currently refer to as artificial intelligence) signals an even more profound transformation for human decision-making and future warfare. The rapid pace of change—one of the key themes of this book—is being driven by the proliferation of big data exploitation and artificial intelligence. At the same time, these tools (and they are just tools) offer humans the capacity to use their biological cognitive skills in new, different, and more meaningful ways.

The result of the partnering of human cognitive abilities with artificial intelligence will be faster and better quality of military decision-making at every level. We have only just started thinking about the impact this will have on our people and on the training, education, and other development mechanisms employed by military institutions. We need to move much more quickly in this endeavor.

Therefore, another measure of tactical effectiveness in twenty-first-century military institutions will be the capacity to sustain an effective and adaptive training and education system that is outcomes-focused, builds cohesive teams, and constructs effective human-machine capability that can achieve tactical success.

### Systemic Integration of Technologies

We have already explored how new and disruptive technologies are changing the character of war. Better and more secure communications, more precise and longer-range weapons, unmanned systems with greater persistence, an enhanced capacity to generate global influence through the Internet and social media, and new forms of artificial intelligence and biotechnology all have an influence. With some exceptions, development of these technologies has taken place outside of military institutions. They must, however, be absorbed into military organizations and combined with new ideas and organizations to be truly effective.

There is some basis in fact for the common perception that military institutions are resistant to change. Records of Western forces at the start of the two world wars, as well as in Vietnam and Iraq, illustrate how slow military organizations can sometimes be to change. However, the historical record of military innovation also demonstrates that military organizations, with the right incentives, can be very creative and highly adaptive. The adoption of aircraft, radios, armored vehicles, submarines, computers, GPS, space-based sensors, and information operations are all evidence of the inherent capacity of military organizations to evolve. Each has had a profound impact on tactics and the conduct of tactical operations.

New technologies alone are not decisive; they must be combined with new forms of thinking and new ways of organizing tactical units to have meaningful impact. This "trinity" of technology, thinking, and organizing is not just important to the continued enhancement of tactical capacity. The process of developing new ideas and organizations that can absorb new technologies also drives the cultural change that is essential to progressive and adaptive military institutions. Therefore, the final element of tactical effectiveness in the twenty-first century is the capacity of a military institution to possess a systemic approach to integrating new and evolved technologies into organizations in concert with new tactics and tactical organizations.

———

In many contemporary militaries, the competition between the development of broader management, governance, and interagency skills and the development of tactical excellence is tremendous. The demands of joint operations, unit administration, and governance and the myriad of secondary responsibilities forced on our junior officers prevent all but the most disciplined from building anything other than average tactical acumen. But tactics and tactical excellence matter. They provide the foundational knowledge for every military leader. Importantly, this level of war has been highly consequential in past conflicts and will be in the future. Military leaders must be brilliant at this basic aspect of the military art and science if they are to have any chance of success in the competitive environment of the twenty-first century.

## ADAPTATION: A MULTIPLIER OF EFFECTIVENESS

The development of military effectiveness at the strategic, operational, and tactical levels should provide well-resourced, cohesive military institutions that are integrated within a national schema to produce effective, strategically coherent outcomes. However, there is one final element of twenty-first-century institutions and ideas that is vital. This is the capacity to change quickly and remain effective in a rapidly changing geopolitical and technological environment.

The breadth of change, and the speed at which it is occurring, must be considered to ensure all levels of military institutions are open to opportunities and resistant to the effects of surprise. Regardless of industry, the generation of a competitive advantage in the "era of accelerations" is becoming more difficult. When an advantage is generated, it is likely to be more fleeting than in previous eras. Rita McGrath has written that we now exist in an era of transient advantage and that successful institutions must spark continuous change and avoid the rigidity that leads to failure.[71] It is through this lens of constantly evolving sources of advantage that nations will need to pursue approaches that harness all aspects of national capacity.

To retain relevance and remain in the vanguard of best practices, military organizations will need enhanced mechanisms for environmental scanning and adaptation. National security practitioners and military leaders must remain at the forefront of understanding the various elements of change and continuity in their environment. Generating diverse options within an ongoing strategic dialogue between agencies, industries, and nations will be critical to sustaining a strategic edge in the twenty-first century. As Peter Schwartz notes in his classic examination of strategic planning, *The Art of the Long View*, "Resilient companies continually hold strategic conversations about the future."[72]

As such, twenty-first-century military institutions and ideas must be designed to win the adaptation battle at the strategic, operational, and tactical levels. Military institutions must be capable of anticipating surprise at the strategic and tactical levels and, if surprised, possess the resilience to survive and adapt. Concurrently, they must support the adaptation of friends and interfere with the adaptive processes of competitors and potential adversaries.

It is impossible for military and other national security planners to anticipate every eventuality. There are too many scenarios and human interactions to accurately predict all the forms of future conflict and competition.

Consequently, a key virtue for effective defense planning has to be adaptability to unexpected events and anxieties.[73] But what is adaptivity, and how does it apply to military institutions?

The earliest work related to investigating adaptation was related to evolution in biological sciences. However, not until Charles Darwin produced his theory of evolution was a causal mechanism developed to account for evolutionary changes.[74] Darwin sought to explain how new species emerge and how others vanish, as well as why existing life forms have the characteristics that they do and why they functioned to assist organisms to achieve specific tasks.[75] This was his theory of natural selection.

Variation was one of the three vital ingredients of natural selection Darwin proposed.[76] He found that variation was an essential part of evolution (and therefore adaptation). However, only some variations are passed down (inherited), and it is these variations that play a role in the evolutionary process. Variation, however, provides the raw material for adaptation. The key is to ensure that a system, in seeking to retain its ability to succeed, balances uniformity (or the retention of lessons learned) and variety (the basis of adaptive capacity).[77] This is critical in military institutions that tend toward conservatism and conformity. Military organizations must be able to generate diverse views to test old ideas and build new warfighting concepts and strategies.

In recent decades, adaptation research has moved beyond the theory of evolution and is now applied in a range of scientific endeavors. Darwin's work has been influential in the examination of how societies organize and how businesses and other collectives might enhance their chances of success in changing environments. Through this work, it has become evident that adaptation underpins learning, development of societies, organizations, and cultures, and solving complex problems by humans.

Adaptation, and the ability of an individual and institution to adjust to their environment, however, does not guarantee success.[78] Systems have demonstrated adaptive capacity but have still suffered partial or catastrophic failure. For example, the U.S. Army, which demonstrated the capacity to adapt to the Cold War central European front and use the tactics developed there successfully against Iraq in 1991, then failed to quickly recognize changed circumstances after capturing Baghdad in 2003.

Therefore, it might be reasoned that a military institution's possession of adaptive capacity alone is not sufficient for continued success in a changing environment. Other parameters are likely to influence the success of a system. In particular, three key measures of success are required: the speed at which a system can "move" in the fitness landscape to more suitable regions; the ability to stabilize and protect useful properties of the system; and the ability to modify the surrounding environment to increase the system's chances of success.[79]

The competitive learning environment of contemporary strategic competition and military operations brings to the fore our need to fully develop and exploit the adaptive processes of military institutions. Whether it is the Taliban (on either side of the Durand line), insurgents in the southern Philippines, or an assertive China, many nations are currently facing adversaries who have demonstrated the capacity to adapt in short timeframes. Western military institutions must also be able to adapt rapidly at the tactical, operational, and strategic levels to remain relevant and capable in the twenty-first century.

Adaptation occurs not only at the individual level, but also at multiple scales within a nation and within a military organization. The key elements of adaptation in a military institution are as follows:

- Build and sustain environmental awareness: Institutions must have the capacity to gain and sustain environmental awareness of the system (geopolitics, national policy, populations, technologies, and relationships) in which the military institution exists and seeks to be successful.
- Develop a view of what is likely to succeed in that environment: This is an institution's views of fitness for that environment, from the tactical to the strategic level. Through the development of this view of what is likely to result in success and the extensive testing of such views, military organizations can then try different approaches to test and validate those notions of fitness.
- Make changes and learn from those changes: Institutions need to be able to make changes (on different timescales and at different organizational levels) based on environmental understanding and notions of fitness.

- Retain and share knowledge: Institutions need the capacity to retain and encode in themselves and in individuals the useful information that improves their chances of success (corporate knowledge). For military institutions, this includes the ability to collect and absorb lessons, disseminate the implications of these lessons (doctrine, new tactics and strategies, evolved training and education, leading change, etc.) and continue to learn based on the interaction of the institution with its environment.

- Measure success and failure of engagement with the environment: This is the capacity of an institution to gauge its actions in moving toward this definition of fitness, which leads to further change in institutional and individual actions, objectives, and notions of suitability.

The exploration of adaptation has resulted in the development of a range of concepts that underpin the understanding of how adaptation occurs and how it can be applied. One of these is the concept of adaptive cycles, which are often defined through the lens of the field being examined.

In military literature, the best-known adaptive cycle is Col. John Boyd's OODA (observe-orient-decide-act) loop. Boyd's fascination with gaining advantage through reacting and maneuvering faster than an opponent was to "constitute the basis for nearly everything he thought and did later."[80] Boyd's 1976 paper "Destruction and Creation" synthesized his ideas and theories to that date.[81] In his later "Patterns of Conflict" briefings, he would seek to expound his ideas to military audiences. In Boyd's 1977 "Patterns of Conflict: Warp X" briefing, the beginnings of the OODA loop philosophy begin to emerge. Boyd continued to develop this thesis until the more mature OODA loop concept appeared in its current form in a 1978 briefing entitled "Patterns of Conflict: Warp XII."[82]

Boyd's work has had significant influence in the military institutions of the United States and beyond. It provided a useful conceptual framework for those who wished to move from methodical approaches of warfare to more maneuverist ways of operating.[83] By the late 1980s, Boyd's concepts received advocacy from theorist William Lind and U.S. Marine Corps General Al Gray. The advocacy by Lind and Gray, widespread debate inside and outside the Corps, and the ideas of Boyd finally coalesced in the publication of *Warfighting* in 1989.[84] It not only contained a new way to think about winning in war; it also nurtured within the U.S. Marine Corps an institutional approach to adaptation.

In a complex environment, the goalposts for success will also be different depending on the scale being examined and the relevant timeline. This complicates the matter and means that measures of success must be designed around the different levels of war as well as different timeframes. As part of an adaptive approach, effective organizational adaptation will rely on a willingness by military leaders to adapt those measures of success, for different scales and timeframes, as the surrounding environment changes.

The implementation of an adaptive approach in an institution also has a corollary: if we can be disciplined enough to adapt at all levels in a cohesive and efficient way, we might also understand how a competitor or adversary adapts. By better understanding how organizations adapt, we may be able to more precisely target those areas of other organizations that we might want to influence or interrupt. Understanding adaptive mechanisms and complex systems, our future military institutions can more efficiently and effectively shape and influence an adversary's ability to adapt. Returning to Frank Hoffman's thesis on adaptation discussed earlier in this chapter, we should be able to interfere with the organizational learning capacity of an adversary. I describe this as *counteradaptation*.[85]

We should therefore focus as much on influencing enemy adaptive measures as we do on fostering our own. An approach to do so must be founded on a theoretical basis of adaptation and complex adaptive systems. In doing so, we should identify the level and timeframe over which we seek influence. It is possible that countering the adaptation of an adversary may see us influencing them at many different levels concurrently, with objectives that have different timeframes.

## A CONCEPT FOR COUNTERADAPTATION

In a 2007 article for the *Australian Army Journal*, I described counteradaptation as "the logical extension of the military counter-reconnaissance battle . . . which seeks to deny information to the enemy of friendly strengths, dispositions, and intentions. Counteradaptation operations aim to enhance friendly capacity to adapt, through the deliberate degradation of adversary adaptive mechanisms."[86]

In their 1990 book *Military Misfortunes*, Eliot Cohen and John Gooch explored significant military failures over the past one hundred years, producing failure matrices that identify the critical pathways to misfortune and failure.[87] In seeking to adopt a more systemic approach to their analysis of failure, Cohen and Gooch defined the three types of errors that can result in either simple or complex failure: failure to learn, failure to anticipate, and failure to adapt.[88] Counteradaptation seeks to induce this *failure to adapt*— or at least prevent effective change—in our adversaries.

Counteradaptation should aim to enhance the relative success and adaptive capacity of friendly elements by influencing and degrading the adversary's adaptive processes. No single action or process alone is likely to sufficiently influence any adversary or group of actors we seek to manipulate. A range of actions will be required, each with desired and measurable outcomes that lead to the realization of an overall objective or strategy. Drawing together the threads of our knowledge in adaptation, complex adaptive systems, and adaptive processes, the remainder of this chapter examines what a concept of counteradaptation may comprise.

Counteradaptation would need to focus on attacking the five elements of adaptation discussed earlier. Table 3.1 compares the link between adaptation theory and the concept for counteradaptation.

Counteradaptation operations should seek to deny an adversary the capacity to effectively adapt to a friendly force's strategy, presence, and activities. This will reduce the enemy's range of options against friendly forces, as well as degrade the enemy's fitness for operations and their capacity to influence friendly operations.[89] Further, counteradaptation operations should also decrease friendly predictability during the conduct of military operations. The five elements of counteradaptation are briefly explored below.

### *Degrade Adversary Environmental Awareness*
All adaptive activity is based, on some level, on response to the environment. The aim therefore is to ensure that the adversary has a qualitatively poorer

Table 3.1. Adaptation and Counteradaptation

| Elements of Adaptation | Elements of Counteradaptation |
|---|---|
| Build and sustain environmental awareness | Degrade adversary environmental awareness |
| Develop a view of what is likely to succeed in that environment | Influence (and corrupt) notions of fitness for the environment |
| Make changes and learn from them | Shape, influence, and corrupt change mechanisms |
| Retain knowledge | Corrupt and destroy sources of organizational lessons and knowledge |
| Measure success and failure of changes and other engagement with the environment | Monitor, degrade, and influence feedback loops |

awareness of the environment than friendly forces do. This will require an appreciation of all the individual and collective actors present, as well as their allegiances and relationships. Sensing the environment (its key elements and the relationships between them) to produce a continuous collection, fusion, analysis, and dissemination of environmental awareness will be critical.

Friendly forces will have to undertake a range of intelligence collection activities that are designed to provide information on the identity and motivations of these other actors. Gaining an understanding of the various interactions of these actors, including any interaction with the adversary (be it positive or negative), will be critical. Improved understanding provides a richer level of detail on the security environment and gives a better indication of how friendly forces are more likely to counter the influence of the adversary in that environment. The degradation of an adversary's capacity, while maximizing friendly capability, is vital. Examples of these methods include deception, signature management, and information operations.

### Influence Notions of Fitness
We must influence how an adversary might exploit the picture they have of the strategic and operational environment. It will then require an understanding of what makes actors "fit" in each environment. From this generic notion

of fitness for a given environment, we can then commence collecting information on competitor or adversary interactions with the environment, and with other actors, that provide the information it uses to ensure its success in that environment. This can be used to identify the high-value elements (be they organizations, technologies, capabilities, or tactics) of an adversary that we would wish to influence, degrade, or interrupt. In this process, it is also critical to ensure the adversary is denied recognition of any patterns that friendly forces set. The key to this is military institutions recognizing any patterns set and making changes before the adversary identifies and exploits them. This will require the capacity to mislead an adversary's assessments of friendly capacity and to protect the capabilities of important military systems.

### Shape, Influence, and Corrupt Change Mechanisms

We must induce (or reinforce) in our competitors and adversaries what Peter Senge has called an organizational learning disability.[90] The aim is to degrade an adversary's capacity to learn effective lessons from their understanding of the environment and their actions within that environment. The capacity to make changes (at different timescales and organizational levels) is based on environmental understanding and notions of fitness. Counteradaptation operations seek to influence, interrupt, or degrade an adversary's adaptive mechanisms. It would influence and interrupt the adaptation cycle of those who seek to prevent friendly influence.[91] This is a good framework upon which to base counteradaptation operations because it is a natural approach to adaptation that any adaptive element would need to possess to be able to survive and influence others within a complex adaptive system.

### Corrupt and Destroy Sources of Corporate Lessons and Knowledge

Corporate knowledge is the capacity to retain and encode in ourselves useful information that improves our success. A competitor's ability to adapt to friendly force operations is reliant on their capability to collect and then disseminate information about friendly activities. Interfering with the enemy's ability to learn more about friendly operations and denying their capacity to share what they do learn are essential to counteradaptation operations. Also known as the counter-reconnaissance battle, friendly forces must undertake an array of active and passive measures to degrade the situational awareness of those who would seek to deny us achieving our mission.

### *Monitor, Degrade, and Influence Feedback Loops*

Measures of effectiveness are used explicitly and implicitly in the military, industry, academia, and other endeavors. The ability to measure the success and failure in moving toward definition of fitness, leading to further changes in actions, objectives, and notions of fitness, is one of the key elements of an adaptive organization.[92] We will require an ongoing assessment process with adaptive measures of success and failure to ensure we are moving toward our objectives while influencing our adversary's ability to adapt.

Whether or not we possess an explicit and systemic approach to adaptation, others will. But adversaries will not be the only ones exploiting adaptation at different levels and in different timeframes. Allies and partners from other government agencies (even contractors) will all be cycling through their own adaptive cycles (even if they do not deliberately attempt to do so). We must be adaptive and win the adaptation battle to deal with the other actors within the environments in which we live and operate.

## TWENTY-FIRST-CENTURY MILITARY EFFECTIVENESS

The convergence of change in the global environment, the rapidly evolving technological milieu, and the physical speed of new weapons systems means that military institutions are more likely to be surprised. Military leaders, organizations, and ideas in the coming decades therefore must provide for enhanced resilience in military and national capacity. They must also build the capability for military forces to anticipate surprise at the strategic and tactical levels and, if surprised, be sufficiently resilient to survive and adapt.

Military organizations are never at a steady state. The rapid pace of change means that they will be adapting at each level concurrently and doing so constantly. At the same time, military forces (and the wider national security enterprise) must be continually interfering with the capacity of competitors and adversaries to do the same. This forms an ongoing adaptation battle— and it is one that is conducted at every level of military and national security activities.

War will remain a human endeavor, albeit one continually evolving due to the impacts of new technologies, different warfighting ideas, and geopolitics. And it will demand continued investment in the ideas and institutions that make up the military instrument of nations. But ideas and institutions by themselves are insufficient. There is one more critical element in the generation of military power that will be explored in the following chapter. This element not only is critical but also is the essential foundation of all military activity: people.

# 4

"All right, you guys, let's get squared away here," he said looking in every direction but at us. (This was strange, because Johnny wasn't the least reluctant to make eye contact with death, destiny, or the general himself.) "Okay, you guys, okay, you guys," he repeated, obviously flustered.

A couple of men exchanged quizzical glances. "The skipper is dead. Ack Ack has been killed," Johnny finally blurted out, then looked quickly away from us.

I was stunned and sickened. Throwing my ammo bag down, I turned away from the others, sat on my helmet, and sobbed quietly.

Never in my wildest imagination had I contemplated Captain Haldane's death. We had a steady stream of killed and wounded leaving us, but somehow, I assumed Ack Ack was immortal. Our company commander represented stability and direction in a world of violence, death, and destruction. Now his life had been snuffed out. We felt forlorn and lost. It was the worst grief I endured during the entire war. The intervening years have not lessened it any.

Capt. Andy Haldane wasn't an idol. He was human. But he commanded our individual destinies under the most trying conditions with the utmost compassion. We knew he could never be replaced.

He was the finest Marine officer I ever knew. The loss of many close friends grieved me deeply on Peleliu and Okinawa. But to all of us the loss of our company commander at Peleliu was like losing a parent we depended upon for security—not our physical security, because we knew that was a commodity beyond our reach in combat, but our mental security.[1]

Eugene Sledge described the death of his company commander on Peleliu with a sense of anguish and loss that reaches out to us across the decades since that October day in 1944. Sledge, a young Marine mortarman who went on to become a professor of biology, produced a wartime memoir called *With the Old Breed* that is now read by students in military schools, by academics, and by others seeking to understand the impact of war on people. Historian Paul Fussell has described it as one of the finest memoirs from any war.

For as long as people have fought, they have also recorded their observations and feelings. Since Julius Caesar produced his *Commentaries on the Gallic Wars*, these memoirs have served to communicate to civilians narratives about the conduct of battles and campaigns, the brutality of warfare, and the valor of individuals and units. The profusion of wartime memoirs from the eighteenth century onward provides many insights into the lives of soldiers, marines, sailors, and airmen across centuries of conflict. Perhaps the most important insight is that war, regardless of where or when it was fought and whatever the technologies and strategies employed, is ultimately about people.

People are the foundation of every military capability. They are also at the heart of every form of military advantage. Military personnel must have the mental agility to thrive in ambiguity, to exploit fleeting opportunities and the ability to continuously learn about their ever-changing environment. They also need to embody the requisite adaptive traits to generate advantage in circumstances of rapid change, while mastering the art and science of their profession throughout the information age. These attributes will determine the success or failure of their institutions.

Modern military institutions have developed various ways to teach their people how to use their cognitive and physical talents. These methods permit them to exploit the full potential of tactics, organizations, and military techniques, as well as their weaponry, equipment, and other tools. However, in this new era of competition and conflict, another aspect must be added to military learning—the capacity to effectively partner with machines. Personnel must be inculcated with the ability to work within human-robotic teams in a training environment, in preparation to augment physical ability on the battlefield. This new era will also demand humans partnering with different algorithms to supplement and speed up human cognitive abilities.

The aim of this chapter is to explore how contemporary military institutions might better anticipate the demands of the future environment in order to build relevant, effective, and adaptive military personnel in the twenty-first century. Every citizen who chooses to serve their nation, regardless of the military service they enlist in, is special. Military organizations have an obligation to provide these people with the very best training, equipment, education, and development.

Vitally, military organizations must also provide the best leadership they can, to ensure that military personnel are physically, intellectually, and morally prepared for the rigors of twenty-first-century military service. Good leadership is a core part of effective military institutions. It is also essential for leading the transformation required in military forces in the coming years. As such, while this chapter covers how military institutions must develop their people, its focal point is the development of military leaders.

Martin van Creveld wrote in his classic study of military leadership, *Command in War*, that "the functions of command are eternal."[2] Carl von Clausewitz described the need for able intellects to lead armies in *On War*. He noted that in order to be carried out with any degree of virtuosity, a complex activity calls for leaders with the appropriate gifts of intellect and temperament. He then noted two indispensable qualities: an intellect that even in the darkest hour retains some glimmerings of the inner light that leads to truth, and the courage to follow this faint light wherever it may lead.[3]

Professional, competent, and creative military leaders do not emerge without significant investment by military institutions. The great captains throughout history, such as Alexander, Hannibal, Caesar, Frederick, and Napoleon, were all clearly gifted. However, they were all products of a diverse range of informal and institutional training, mentorship, and education. The complexities of military operations, regardless of the era, mean that military leadership must be taught, practiced, and continually honed by institutions. Neglecting the development of military leaders has proven, throughout time, to have profoundly bad outcomes for military organizations and, sometimes, for their countries.

The intellectual development of military people and their leaders reflects the duality of war; it retains some enduring features but will also continue to evolve. While we may be entering a new era, military institutions cannot arbitrarily discard all the previous learning of the profession of arms. This would expose them to significant future risk of failure. Therefore, the educational systems of military institutions will need to continue reinforcing core aspects of military service, reinforcing the "institutional" over the "occupational" approach to the profession of arms.[4]

Nevertheless, the changing character of warfare means that the process of identifying and developing military leaders will also continue to evolve. Military institutions must prepare their leaders to rapidly adapt to changing situations, new technologies, and different ideas. Yuval Harari describes the imperative for humans to adapt in his book, *Homo Deus*: "In the early twenty-first century, the train of progress is again pulling out of the station—and this will probably be the last train ever to leave the station called Homo Sapiens. Those who miss this train will never get a second chance. In order to get a seat on it you need to understand twenty-first-century technology, and in particular the powers of biotechnology and computer algorithms. . . . Those left behind will face extinction."[5] This is a clarion call for those who lead the intellectual development of military leaders. Military institutions in the twenty-first century must be on the train that Harari describes as it leaves the station.

If the military profession is to effectively contribute to defense, deterrence, and other national security outcomes into the twenty-first century, it must continue to adapt. Moreover, the military profession must be capable of doing whatever societies demand, across a diverse range of domestic and international endeavors. This might mean fighting high-level conventional wars or providing support to civil authorities in responding to pandemics and natural disasters. Importantly, the military profession must support the sovereignty of the nation it serves and continuously generate a competitive edge over adversaries during periods of strategic competition and conflict.

## GENERATING MILITARY ADVANTAGE THROUGH PEOPLE

There is no tabula rasa when designing the future development of military personnel, nor is one desirable. A blank slate would discard all the previous learning of the military institution and necessitate a total reboot of training, education, and development systems. But change is required.

Technological disruption and recent developments in national security affairs are challenging the orthodoxy established in the 1950s with regard to the profession of arms. The analogue, machine-based world that existed when *The Soldier and the State* and *The Professional Soldier* were written has changed considerably. In addition to traditional instruments of war, current and future military leaders must prepare for the application of various nonkinetic instruments that will threaten our national security, in addition to the more traditional military capabilities. Furthermore, they need to be cognizant of the fact that nation-states will no longer have exclusive control over large-scale violence and activities that will impact upon strategic influence.[6]

Military leaders seek to generate advantage over their adversaries. While it might take many forms, advantage historically has been generated in one of five ways: geography, mass, time, technology, and intellectual advantage.[7]

Clever military leaders generate advantage by exploiting geography to ensure they can gain and retain the initiative against an adversary. Geography has long held an important place in how nations and their military forces build a competitive advantage. As Colin Gray wrote, "Geography is the most fundamental of factors which condition national outlooks on security problems and strategy solutions. The influence of geography truly is pervasive, notwithstanding the fact that that influence must vary in detail as technology changes."[8]

For a nation such as the United States, geography has provided physical security for much of its modern history. Likewise, Australia's location and its status as an island, continent, and unified nation have provided physical security, and shaped its defense policy, since the beginning of European settlement in 1788. Nations that share long land borders with other countries, such as Switzerland, China, or Germany, have very different security challenges.

The advantages of good geography are not what they once were. Security through geographic distance has declined in the age of intercontinental missiles, cyber operations, globalized supply chains, and information warfare. The speed of connectivity, long-range sea and air transport capabilities, and the ability of individuals to move at will to any point on the globe mean that geography no longer guarantees sovereignty.[9]

Mass is a second historic source of a military capability edge. Military institutions have long aspired to generate a larger force than an adversary. Whether it is to provide the capacity to concentrate forces and achieve local overmatch at the tactical level, or to exploit in order to provide the requisite scale necessary to operate across different parts of the globe, mass has played a crucial role in historical military success.

The doctrine of massed forces has influenced generations of military leaders in the West. Sun Tzu wrote that "if we are concentrated into a single force while the enemy is fragmented into ten, then we attack him with ten times his strength,"[10] while Clausewitz noted that "in tactics as in strategy, superiority of numbers is the most common element in victory."[11] Antoine de Jomini included mass in his 1838 *The Art of War* as a principle of war. He wrote that "there is one great principle underlying all the operations of war—a principle which must be followed in all good combinations. I. To throw by strategic movements the mass of an army successively upon the decisive points of a theatre. . . . II. To engage fractions of the hostile army with the bulk of one's forces. . . . III. To throw the mass of the forces upon the decisive point."[12]

This idea of deploying a large military force does not just include the number of people in uniform. As the U.S. Civil War and two world wars demonstrated, successful military mobilization also requires efficient mass industrial mobilization. In the first half of the twentieth century, the United States and the Soviet Union developed the capacity to mobilize large numbers of people; ensuring that industry could keep them adequately equipped, fed, and supplied became an important military and national capability.[13] Having large military forces might be useful, given it allows for many concurrent activities. But, as we explored in chapter two, the meaning of "mass" is different in the twenty-first century. The types of conventional mass that may have been decisive in the nineteenth and twentieth centuries will not have the same impact in the twenty-first century.

A third way of generating military advantage is through the clever use of time. In chapter two, we explored the implications of time for military activities and how the current surge in innovation is resulting in faster technological change. Robert Leonhard posits that "we are in the midst of the most exciting revolution in military art and science since the invention of gunpowder. At the center of it all will be a contest for time. Staid principles of industrial age warfare will not avail. Some of the most fundamental notions about military operations will have to be retired. Like successful commanders of the past, tomorrow's warriors must act not by laws, but by minutes."[14]

Being able to effectively use time and deny it to an adversary is one of the great arts of military command. For example, an aggressor might exploit time by deceiving their adversaries about their intent, affording themselves more time to prepare and conduct offensive operations; this denies the victim of the surprise attack time to prepare. Historically, successful military leaders have been able to look beyond numbers and seize the initiative through their clever use of time and their generation of a more effective operational tempo than their enemy.

The fourth method that military forces can use to generate advantage is superior technology. From Greek fire to crossbows, tanks to jet aircraft, the ENIGMA machine to contemporary high-capacity computing, military institutions throughout history have sought to secure a competitive edge by possessing better technology than their adversaries. Advanced technology has provided an edge for Western military forces for generations, but its advantages are significantly less than in the past. Recent publications have described in detail the decline of the technological edge that has been the preserve of Western military institutions for several centuries.[15]

During periods of conflict and strategic competition in the decades ahead, Western military organizations will face significant challenges to these four historic elements of military leverage and influence. Geography is no longer a guarantor for retaining national sovereignty, and Western militaries lack the mass of their potential adversaries. Furthermore, the advantages that were once afforded by more advanced technologies are now declining. Complicating this situation further, when Western nations might generate a technological advantage, it is likely to be only a transient rather than an enduring one.[16]

As a consequence, military organizations must devote more resources to cultivating a fifth source of advantage: the *intellectual edge.* The clever application of military forces within a national approach is built on the best ideas being applied to tactics, operational concepts, strategy, and different organizational constructs. A well-honed intellectual edge can be used to offset the growing military, financial, and information capabilities of potential adversaries.[17] Possessing an intellectual edge over an adversary provides a source of strength. It can also be applied to unify other sources of national strength into a greater whole.

An intellectual edge has two components. The first is individual excellence, or what might otherwise be called professional mastery. The intellectual edge for any individual military leader is the ability to creatively out-think and out-plan potential adversaries. This requires a diverse range of training, education, experience, and talent management. It also demands one's personal dedication to continuous self-learning over a long period of time.[18] Building this intellectual edge requires individuals who can develop their capacity for relationship-building across many different types of military and nonmilitary groups, as well as different nationalities. Having a finely honed intellect is pointless if it cannot be applied to influence others through effective human interaction.

The individuals who dedicate themselves to this pursuit must also be able to leverage the investment in their intellectual development to achieve team-based objectives. This leads to the second component of the intellectual edge, which is *institutional.* A collective, institution-wide intellectual edge stems from an organization's capacity to effectively nurture and exploit the disparate intellectual talents of its individuals to solve complex institutional problems. The challenges of future force design, operational concepts, the integration of kinetic and nonkinetic activities, and personnel development must be the targets of this institutional intellectual edge.[19]

Such an approach requires military institutions to sustain and invest in their recruiting, individual and collective training, and educational activities. The development of military personnel and future leaders through education and training must be conducted within an overall system that nurtures the talents of people within their occupational specialties. In this training and education system, there is also a need to identify the most talented people and assign them to additional challenging intellectual development opportunities.

The institutional intellectual edge has another important component— "institutional learning," which is distinct from (but related to) the institutional talent management, training, and education that it provides to its individual personnel. Achieving effective learning underpins the adaptative capacity of military institutions that was explored in chapter three. Military institutions must possess the essential mechanisms for learning, which include training and education, formal study of lessons learned processes, a means to disseminate lessons, and the mechanisms to adapt ideas and organizations quickly. They must also cultivate an institutional culture that embraces and incentivizes learning at all levels. Further, institutional learning must encourage the observation of, and learning from, the expertise of allies and adversaries.

One of the best case studies of the institutional approach to an intellectual edge is British scholar Aimee Fox's study of the British army in World War I, *Learning to Fight*. Fox notes that "the overall effectiveness of its processes for learning was contingent on several different factors. The army's ethos enhanced its ability to learn and adapt. It also influenced its approach to learning. Out of necessity it was required to prioritize and modify different ways and means for learning. Through a combination of this flexibility and its ethos, the army promoted a culture of innovation across its operational theatres where individuals were given the opportunity to influence institutional behavior."[20]

The relevance of this study of institutional learning in World War I is clear. The British army comprised both professionals and recently drafted personnel, and its operations spanned multiple theaters across the globe in different climes and domains. It was fighting in a conflict that was for the first time making use of many new technologies from the second Industrial Revolution, such as aircraft, wireless communications, rapid-firing small arms weaponry, and large-scale standardized manufacturing. The adoption of these new technologies required the intellectual efforts of humans.

While it is humans who use technology, the impact of technology on the intellectual edge will also be important. The future intellectual edge for individuals and institutions is likely to require cognitive support through human-AI teaming. Described as "System 3" thinking by Frank Hoffman, this new field in the collaborative application of biological and machine intelligence will increasingly be core to the development of the intellectual edge.[21] Excellence in twenty-first-century military endeavors will be driven by institutions that can nurture and apply an intellectual edge, across the spectrum of strategic to tactical activities, with an optimal fusion of biological and artificial intelligence. The focus of the next section is to explore how military institutions might build this edge in the future.

## A FUTURE INTELLECTUAL EDGE

All the great professions move their practitioners through different career stages. Doctors, for example, undergo initial education and then attend medical school before they progress into an internship. After their internship, doctors either enter general practice or a specialty. Regardless of which pathway they choose, continuous development and education will be needed to keep abreast of the changes in their profession. Engineers also follow a career progression model that incorporates initial education, initial experience, and subsequent specialization; these steps are followed by undertaking senior engineering responsibilities and/or senior management functions.[22]

The same framework that supports the progression model within many civil professions is necessary for military personnel and their leaders. However, given the length of a military career and the diversity inherent to the various specialties within defense, it is not possible to describe a singular vision for a twenty-first-century military leader. Nevertheless, we can form a view of what functions this military leader might need to perform within the principal stages of their military career. To identify and scope the key performance criteria within each stage, we will need to draw upon on knowledge of the military profession as well as an understanding of the future security environment.

Military institutions must possess a view of an individual's intellectual journey from their first day in service to their last. Such a continuum provides for an adaptive, future-oriented performance specification for military leaders. For the purposes of this discussion, I have adopted a five-stage approach, similar to a recent Australian Defence Force model.[23] It illustrates how military leaders progress through their careers.

Each of these stages is additive. The individual stages provide a foundation for subsequent growth and development of personnel. This progression through the stages of a leader's professional development is multifaceted, combining experience, interaction with external agencies, and training and education. The desired outcome is to forge these components with the cognitive abilities that underpin critical and creative thinking in planning and the exercise of command.[24]

The five stages also make up the baseline of skills that the twenty-first-century military leader will require, not the entirety of their repertoire. They will need to be experts in at least one domain; currently, the services oversee development early in the career of military personnel. Many others may need to be specialists in areas such as communications, logistics, and engineering. This specialist expertise may also be needed in new areas as technology evolves. Notwithstanding, a staged continuum will form an essential backbone for a system that will nurture military leaders. It represents a manifestation of what Charles Moskos called "an institutional core," providing a unifying approach for generalists and specialists.[25]

### Stage One: Building a Professional Foundation

From initial training, military leaders must establish a foundation for leadership in military institutions. Much of this will be the remit of the individual services—the navy, army, air force, or marines. Most nations retain academies to undertake initial officer training in a single service or domain environment. This is also a critical phase for inculcating the values, attitudes, and behaviors demanded by military organizations and the profession of arms.

In the earliest part of their development, military leaders must learn to contribute to team goals through cooperation and demonstrate the ability to build good team relationships, while displaying high ethical and professional standards and practices. These young leaders will be required to convey ideas within their own teams.

The ability to communicate is one of the most basic yet vital functions of good leadership. Development of this skill must begin at the very start of a military leader's career. However, development of communication skills will be continuous throughout the career. The ability to speak and write in a manner that is understood (and not misunderstood) must be honed through many different experiences. Military institutions play a large part in this process by incentivizing the types of behaviors that develop good communication skills. For example, the ability to speak clearly, confidently, and succinctly is the foundation of developing the capacity to give a good set of orders. Likewise, the ability to write convincingly and produce evidence-based, critically analyzed papers is shaped by institutional incentives such as mentoring, professional journals, and senior advocacy.

One other aspect of developing refined communication skills in military leaders is both urgent and highly necessary: they must be able to speak in plain language. For several generations, military personnel have cloaked themselves in an arcane jargon, full of acronyms. This jargon is nearly impenetrable for outsiders, including those from other government departments related to national security and the elected members of congress or parliament for whom they ultimately work. In an era where joint, interagency, and coalition operations have become the standard, the ability to communicate clear, concise, cohesive, and consistent strategic messages is vital. The removal of exclusionary forms of communication must be regarded as imperative.

At this point in their professional development, military leaders would be expected to display intellectual curiosity. Put simply, they must want to learn. Junior leaders need to be developing a discipline around personal learning and problem-solving, fostered by their leaders, mentors, and institutions. The current surge of change in technology and the follow-on effects in concepts and organizations mean the military leader will need to constantly absorb new knowledge and develop new skill sets. If military institutions are to prove successful in preparing their junior leaders for this new environment, they will need to inculcate the mindset, skills, and disciplines for them to "learn how to learn" in the embryonic stages of their careers.

In the coming decades, a more integrated application of national power may drive many countries to develop joint leader training and development regimes earlier in the careers of their people. Military leaders require an appreciation of the context and roles as an emerging member of the profession of arms. This demands an early introduction to the joint force environment.

The skills described here comprise a foundation for our twenty-first-century military leaders. In many respects, these skills represent a continuity with previous generations of military professionals. That said, one important difference will be the need to be significantly more technologically literate. The military leader of the future must be able to operate within a technologically complex environment, where the understanding and application of advanced information, weapons, and biological technologies represent "business as usual." They will need to learn to work effectively within human-machine teams. Commanding and collaborating with robotic systems and making decisions supported by various artificial intelligence and decision support structures will underpin their development of tactical mastery.[26]

A final, and indeed indispensable, element of the professional foundation for military leaders is instilling in them an understanding of basic military behaviors and discipline. These include the important aspects of any professional military force that includes military discipline, timeliness, physical and mental fitness, conduct, and behavior. It encompasses the values of military institutions and enforcing ethical behavior as well. The responsibility for overseeing these basics of conduct in a professional military institution rests with both officers and noncommissioned officers. When the basics of military discipline are not enforced and lax standards take hold, more serious breaches of conduct can occur. We have seen this throughout history and, more recently, in the conflicts in Iraq and Afghanistan. Dedication to brilliance at the basics of the military profession is an essential foundation for development of more advanced skills in tactics, the operational art, and strategy.

## *Stage Two: Developing Tactical Mastery*

In the next stage of a military leaders' professional growth, they must continue building their foundation for leadership. Much of this development will occur by building tactical acumen within their own service and specialization. However, they must also begin to understand the purpose, procedures, and protocols in the conduct of joint integrated operations.

The most vital skill is leadership. The people appointed to command must be able to lead and influence. To be effective, they must develop the command presence to convince others to do very difficult things in trying circumstances, actions that only military organizations are expected do. These aspects of leadership are inculcated and practiced during the foundational stages of the military profession.

However, it is during this next stage of the military leaders' career when they will develop a more sophisticated and nuanced approach to leadership. In doing so, they must also hone their capacity to trust those they lead. As Martin van Crevald has written, "Historically speaking, those have been most successful which did not turn their troops into automatons, did not attempt to control everything from the top, and allowed subordinate commanders considerable latitude has been abundantly demonstrated."[27] Perhaps Field Marshal Bernard Montgomery best expressed this when he wrote in *The Path to Leadership* that "it is human beings, men and women, who are the factors of reality in the world in which we live; once you can win the hearts and the respect of those who work for you, the greatest achievements become possible."[28]

At this stage of their career, military leaders will need to possess sufficient knowledge and experience to be able to exercise mission orders or mission command. Mission command is drawn from the German concept of *Auftragstaktik*, in which subordinate leaders at various levels execute decentralized operations based on mission-type orders. These mission orders provide subordinate leaders with missions but provide flexibility in their execution.[29] The mission command approach to leadership will continue to be effective into the twenty-first century. By engaging human initiative and creativity, this command approach relies heavily on a clear expression of the superior commander's intent, facilitating rapid, concurrent, and decentralized activities.

For this mindset to permeate the layers of a hierarchal command structure, a "zero-defect" culture must be avoided. Such a mentality stifles the growth of young leaders, preventing them from learning and developing to their full potential. Further, it poisons their minds with ideals and exemplars of the worst possible impressions of what "right" looks like with respect to military leadership. An appropriately tuned command and training climate is a critical enabler for building military leaders who can learn by making mistakes and can execute mission command in the more technological and demanding decades ahead. This command climate must be underpinned by clear expectations of what is unacceptable failure and what is acceptable failure.

Junior leaders at some point in their professional development must be given experience in working within human-machine teams and applying artificial intelligence in decision support. Military institutions will have greatly increased numbers of robotic systems in the coming years, and future leaders must be proficient in applying autonomous systems in combat, logistics, reconnaissance, and disaster assistance missions.

Military institutions will expect their military leaders to have a deeper comprehension of organizational purpose and direction at this point in their professional growth path. There will also be higher expectations of them to not only support the organization internally but to also actively promote its purpose and direction to external agencies. In doing so, they should be able to achieve their assigned missions and objectives by making the best use of their team's and their personal professional, technical, and social mastery. This capacity to see the bigger picture becomes an important discriminator in talent management and judging those able to cope with increased responsibilities at higher ranks.

By this stage, the twenty-first-century military leader should have developed the capacity to employ creative thinking to produce novel ideas and apply critical thinking techniques. Given the rapidly evolving capacity of computing power and artificial intelligence, and the potential for computers to achieve what Nick Bostrum has titled super-intelligence, creativity and ingenuity will be critical skills for future officers.[30] This creativity will both underpin their leadership and planning capabilities and guide their self-learning activities to build and sustain excellence and relevance in their profession.

### Stage Three: The Operational Artist

For the first decade of military leaders' careers, military institutions focus their development on the core skills in their chosen service and a foundational understanding of the profession of arms. Around the ten-year milestone, however, they should begin to transition from tactically focused activities to building excellence in higher level joint activities. Building on the tactical mastery achieved by this point, the military leader should progress to developing the ability for thinking and working at the operational level.

Western doctrine has largely accepted that between military strategy (the ends, ways, and means of military activities) and tactical execution lies an area of military endeavor that links strategy and tactics. Introduced into U.S. doctrine in the 1980s, it has since been included in the doctrine of its allies and partners. As one U.S. doctrinal publication notes, the "three levels of warfare—strategic, operational, and tactical—model the relationship between national objectives and tactical actions."[31] U.S. military doctrine defines the operational level and also provides insights into how to think about joint operations through *operational art*: "The operational level of warfare links the tactical employment of forces to national strategic objectives. . . . Commanders use operational art to determine how, when, where, and for what purpose military forces will be employed, to influence the adversary's disposition before combat, to deter adversaries from supporting enemy activities, and to assure our multinational partners achieve operational and strategic objectives."[32]

Contemporary doctrine therefore provides useful insights into the context for preparation of leaders at this point in their career. This is a time where future leaders must consolidate the experiential learning provided by their service and gain initial experiences in joint and institutional appointments. Importantly, military leaders should now be focused on the educational rather than the vocational aspects of their professional development. They are at a point where a joint education continuum nurtures the capacity for more strategic understanding while building expertise in the operational art.

The military leader must therefore be able to place operational and institutional problems within a larger strategic context. Doing so provides a deeper understanding of "purpose" for the individuals and the teams they lead. Concurrently, they must fully appreciate the limitations imposed by available resources, time, and technologies and ensure that the solutions to problems are developed with these constraints in mind.

At this point in their professional development, military leaders must also look to expand their knowledge of competition and war. This will enable them to actively contribute to the framing of national military strategy and operational planning. They should be introduced to the institutional processes for military capability development to build their appreciation of how to operationalize emerging technologies. In doing so, they should now be undertaking studies alongside civilians as well as officers from across the different domains of military service. This collaborative education with defense civilians not only reinforces the civil control of the military in democratic nations but also ensures that military leaders build an awareness that force is just one instrument of national power. It ensures that civil-military collaboration is fostered early in careers of military leaders and provides the foundation for understanding the policy and strategy context for operational planning.

Our military leaders should now appreciate the interaction of humans and technology, building their understanding of both the art and the science of war. The capacity of military leaders to think about and communicate new approaches, while mentoring their subordinates to think creatively, will be a critical skill set. They develop the ability to apply this in the day-to-day execution of their duties and provision of guidance to subordinates. Also, they must develop the discipline of using this knowledge to contribute to improving their institution. The capacity to move rapidly in the physical and intellectual spheres to improve an organization is a key element of an institution's culture.[33] It is something that can only be fostered by those who connect the tactical elements of the military organization with strategic leadership. In building their capacity to do so, military leaders move beyond being experts in their jobs to developing into stewards of their institution and their profession.

### *Stage Four: The Nascent Strategist*

At this point of their professional journey, military leaders should have commenced preparation for institutionally important senior command and staff appointments. They are developing the capacity to understand and apply military power in support of national objectives. They have moved beyond the realm of tactics and must now be investing in their capacity to think strategically, to understand policy, and to use influence.

By this stage, military leaders must have developed baseline strategic thinking skills so they can appreciate the employment of the national instruments of power. This will help them to take part in developing national military strategy, policy, and operational plans. These facets must align with the direction and policy of the elected government, while at the same time, they need to be mindful of changes in the strategic environment and incorporate emerging technological trends.

Military leaders at this point in their careers are expected to develop strategic direction and long-term plans for their organization. They would be doing so to create a shared sense of institutional purpose and priorities in an environment of ongoing change and uncertainty. Further, they must be able to build and sustain interagency, intergovernmental, and multinational relationships in support of their institution. These not only underpin military and national security planning but also provide foundational networks as they move into more senior appointments.

Two final requirements round out the performance needs of the nascent strategist: organizational change and technological literacy. First, military leaders should have developed a working understanding of organizational theory, institutional cultures, adaptation theory, and change management methodologies. They will require the cognitive capacity and mental agility to appreciate when external environmental demands are changing, to know and understand the capacity for the military to adapt and, moreover, to encapsulate the ability to lead that change at the decisive point. These will all prove to be critical capabilities that will be expected from our future leaders. Further, this will not be a "once in a career" event or even just an annual one. We all will exist in a milieu where change, decision-making, and institutional evolution will occur at a rapid pace. The understanding of key concepts in institutional culture and organizational adaptation is an important part of the knowledge sets of military leaders.

As part of their role as agents of organizational change, military leaders must also be capable of acting as facilitators for change and innovation. It is not a simple skill to master. Organizational culture and deeply ingrained personal and institutional habits can obstruct even the most creative and energetic innovators. Much of this innovation emerges from the lower levels of military institutions, where military personnel face problems that must be solved. These innovators, however, can quickly run up against difficulties in either their chain of command or their organizational culture. Therefore, they need to be supported by institutional "agents of change"—people who understand the need to advocate for innovation and nurture an environment where innovation and creativity are encouraged and incentivized. Military institutions, therefore, need to educate future leaders to embrace this role of innovation supporter and nurturer.

The second capability required by twenty-first-century military leaders is technological literacy—the ability to understand (and keep up to date in) the broad range of existing, evolving, and new technologies that are either used by or have an impact on the military. Military leaders will work within a human-machine environment that extends from tactical human-robotic operations through to strategic decision support by advanced algorithms. The developing strategic leader must appreciate the challenges and opportunities of employing these technologies and ensure that decision support provided by artificial intelligence has sufficient quality control mechanisms. Additionally, at this career stage, the future leader will be making investment decisions in new and emerging technologies. These must be informed by constant attention and disciplined commitment to technological education, literacy, and ethics.

The nascent strategist is an institutional and national asset, someone who can develop better approaches to the strategic challenges military organizations will face in the coming decades. Doing so demands commitment and self-discipline. Most of all, it requires intellectual humility so that they can understand that even at the most senior of ranks, ongoing learning is a core part of being a military leader and a member of the profession of arms.

### Stage Five: The National Security Leader

The culmination of military leaders' careers is their appointment to command their service or a joint military organization. These appointments represent the pinnacle of their institution and the profession of arms. Leaders at this level will need to build upon their multi-decade investment in their intellectual capacity. They must be able to influence and implement national strategy holistically. This is done by orchestrating all instruments of national power in a coherent, synchronized plan to achieve national objectives in times of peace, crisis, and war.

Moreover, these national security leaders will be responsible for the design and maintenance of an operationally effective joint force. In doing so, they must be able to ensure that their organization can align current and future operational concepts and decisions with available technologies. They must also be able to ensure that these concepts incorporate and exploit a diverse range of military capabilities including intelligence, logistics, cyber and communications, training and education, as well as command and control.[34]

By this stage, the most senior leaders must also be capable of setting the strategic direction for the service they lead within a wider defense organization. At the same time, these leaders must be exerting influence within the national defense enterprise to gain resources. They are leaders and influencers who create a shared sense of purpose across governmental, interagency, and multinational organizations to achieve national security outcomes. As a consequence, military institutions need several other strategic-level competencies in these most senior military leaders.

First, they should be capable of providing advice to political leaders on the alignment of strategic goals and actions, commensurate with the size, structure, and capabilities of the military forces of their nation. This is to ensure that strategic documents, such as declared military strategies, contain logical and effective national military objectives. In doing so, they must also possess a well-honed appreciation that strategy making and execution are inherently messy and often unsatisfying. As Beatrice Heuser notes, "If we review the recorded wars of human history, it is rarely, if ever, possible to point to any precise dividing line between politics and the use of armed force as its tool. Strategy in practice rarely follows the precepts laid out by strategic thinkers. Those thinking war will always hope for it to be neater than it turns out to be in practice."[35] An appreciation of this lack of neatness, while also remaining deeply knowledgeable about the theories of war and strategy, will remain an important part of the future military leaders' repertoire.

National security leaders will therefore need to embody a high degree of personal resilience and adaptability, as they will be engaged in continuously reforming and streamlining organizational structures to meet the challenges of changing political environments. As William Rapp has written, "The onus is on military leaders to cross the divide to meet civilian policy makers on their turf, rather than expecting civilian leaders to provide the military clear autonomy in the development and execution of strategy."[36]

These most senior of military leaders need to ensure that the strategic goals for the military organization they lead have objectives that are consistent with their nation's industrial and technological base. This will require them to have a deep understanding of contemporary defense industries and the principles of mobilization in the event of a major conflict. A military leader must be able to integrate their strategic objectives with those of their allies and other security partners or alternatively, the ability to persuade them in concert with their civilian leaders to adopt consistent strategies.[37]

There is one final quality required by these senior military leaders. It might seem obvious, but it absolutely must be restated: They must know how to win. After all, warfare is a competition. It is about humans seeking to impose their will upon other humans through a combination of influence and violence, always in different proportions depending on the circumstances—the importance of the goals sought, the technology available, and the era the conflict occurs in. As Clausewitz reminds us, "To overcome the enemy or disarm him—call it what you will—must always be the aim of warfare."[38]

There is a substantial challenge to defining what winning is.[39] Military leaders must understand what strategic success looks like, well beyond success in military activities alone. They will work in an environment of constant competition, between different states and many different levels.

Defining success—"victory"—in these circumstances is every bit as important. If senior leaders, both military and civilian, can understand with clarity
what they are striving to achieve, develop a clear concept for victory, and
clearly articulate this message to their people and allies in a simple and compelling way, they have a very good chance of success.

Our military leaders must capitalize on their professional education,
experience, and intuition in shaping their critical thinking skills in assessing strategic options for their government. They must be champions of the
profession of arms, the ultimate steward. They lead their institution in its
national security and military endeavors and must therefore be an exemplar
for their personal commitment to lifelong learning and a model for "what
right looks like" in mastering the profession of arms.

## BUILDING THE FUTURE INTELLECTUAL EDGE

The path to building military personnel who can successfully lead in the
twenty-first century is a multistage journey that at times can be arduous. It is
complicated in peacetime by what Williamson Murray has described as the
"privileging of certainty and low-risk behavior, instead of preparing themselves widely and deeply for the uncertain, high-risk endeavor that is war."[40]
Changes in the strategic environment and novel, disruptive technologies will
have a significant impact in the intellectual development of military leaders.
But above all else, they must appreciate the centrality of human agency in
warfare and in the current global system dominated by strategic competition and rapid change. This demands a continuous process of learning and
development.

Continuous learning represents a significant undertaking for military
institutions and demands ongoing investment in facilities, staff, and other
resources. If military organizations are to develop effective military leaders who can avoid "slovenly thinking,"[41] they must continue to develop
the mechanisms to prevent their organizations from falling into this trap.
There are many aspects to the design for building our military leaders
and ensuring they possess an intellectual edge. It will require a systemic
approach, with many interlinked initiatives that encompass goal-setting,

incentives, talent management, and promotion systems that are in alignment with continuous learning through training, education, experience, and networking.

In constructing this system, I believe that institutional leaders will need to focus on and invest in a trinity of initiatives: strategic direction and integration, new-era learning initiatives, and technological requirements.[42]

### Strategic Direction and Integration

The training, education, and experience of military leaders—past, present, and future—occur largely within military institutions. Regardless of whether these are domain-specific organizations such as the navy, army, marines, or air force, or joint institutions, institutions will play a significant function in developing our military people and their leaders. This development must have strategic purpose and a transparent design and become more integrated. The key elements of strategic direction and integration to develop our military people are described in the following pages.

### A Strategic Design

Humans are a core element of the military system. The development of military personnel, through recruiting, education, training, experience, talent management, and other mechanisms, provides the essential "software" that closely binds an integrated force together.[43] In the military context, this software is built by the sharing of ideas, processes, and culture within our organizations. Our military leaders will be spearheading this initiative. Therefore, an institutionally endorsed view of a military leader is required, and this should be part of a broader view of military capability.

Military institutions therefore need a strategic design for how they might develop the intellectual edge in their people. A strategic design should draw from existing institutional cultures and our knowledge of future challenges to build the range of institutional, educational, and technological elements of the professional military education ecosystem that will incorporate all the delivery systems that are fit for the kind of personnel that are entering military institutions. Moreover, such a design requires a mix of residential and nonresidential mechanisms, as well as online, unit-based activities, networking capabilities, and self-learning hubs and incentives.

However, before commencing the design phase, a diagnosis is needed. It might take into account the current situation, known strengths and weaknesses, anticipated future challenges, workforce structures, tasks, and resources. This diagnostic activity is an important part of building the strategic design for developing the intellectual edge in our military leaders. Indeed, the diagnosis is a critical element of any strategy-making.

Getting the question right must be the first step in any problem-solving activity. This diagnostic activity therefore should define or explain the nature of the challenge. Strategist Richard Rumelt has called the diagnosis one of the three components of all good strategies: "At a minimum, a diagnosis names or classifies the situation, linking facts into patterns and suggesting that more attention be paid to some issues and less to others. An especially insightful diagnosis can transform one's view of the situation, bringing radically different perspective to bear."[44]

Once the diagnosis is complete—and it will vary between institutions in different nations and with different strategic cultures—only then can an institution produce a strategic design. This design will compromise a "strategic learning ecosystem" and must be augmented with environmental scanning capabilities and adaptation processes to ensure the various delivery mechanisms are weighted appropriately. This weighting will depend on the requirements of different services and specialties, geographic dispersion, time constraints, as well as the competing demands placed on individuals at different stages in their careers.

An important part of this strategic design will be a vision of what the military institution seeks to achieve in developing its people. One of the most useful explorations of vision is a 1990 RAND report that explored the transition of military institutions at the end of the Cold War. It described the essence of a vision in a military organization: "The sense of identity and purpose found in an organizational vision provides the members of the organization with more than they can find in corporate strategies or long-range plans. A vision provides the essential intellectual foundations for interpreting the past, deciding what to do in the present, and facing the future. It conveys clearly, unambiguously and uniquely what the organization is and what it is all about."[45]

A strategic design, and its inherent vision, should provide the basis for investing in reinvigorated intellectual development across the military enterprise. It should lay down key markers in the development of military leaders and allocate the necessary resources to realize these objectives. This strategy should also ensure the alignment of civil and military education, career, and talent spotting/management, especially at senior levels, to foster a more collaborative civil-military national security environment.

Time and resources, however, will constrain the achievement of a strategic design; the strategic environment will continue to evolve quickly. Therefore, the strategic vision and design must also contain priorities that are revisited regularly. Periodic reviews will ensure that the balance of investment in the different elements of the ecosystem will continue to generate the best return on investment for military institutions and military personnel—especially leaders, who must be prepared for the circumstances and environments that they are most likely to encounter.

## Joint by Design, Not by Convenience

The earlier chapters of this book described the requirement for a more integrated approach to working in the national security environment. This means that the development of military leaders should be underpinned by the nurturing of a more joint, integrated mindset earlier in their career. Consequently, military institutions might better develop the performance specification for their military leaders as *joint by design* rather than *joint-qualified*.

In most military organizations, initial officer training is conducted in single services. The organizations subsequently seek to retrofit a joint and more integrated mindset in these officers some years (or even more than a decade) after they complete their initial training and indoctrination. Given the emerging demands for a more collaborative mindset, is there not a better approach? The Singaporean military has a different officer training model in which a "common training" period for army, navy, and air force officer cadets occurs at the beginning of their service, prior to their transitioning to single-service training.[46] This may be a model worthy of consideration if military institutions are to build the more collaborative and joint military leaders needed to meet strategic challenges.

## Strategic Engagement

Learning activities require interaction between different individuals and a variety of institutions. Military institutions explicitly expect their people to learn from each other during training and education activities. During the coalition warfare in Europe and the Pacific between 1939 and 1945, the various Allied nations often (but not always) shared lessons between theaters and military organizations. The U.S. Army studied the 1973 Yom Kippur war and fundamentally changed itself as a result. Furthermore, almost every military institution in the world studied the lessons of the 1991 and 2003 invasions of Iraq. Many nations have openly shared tactical, operational, and strategic lessons from the wars in the post-9/11 era.

Learning from the mistakes and successes of other military organizations is part of the culture of many military forces. Frequent engagement between like-minded military institutions, services, and nations must continue to evolve and embrace a culture that promotes the rapid sharing of high-quality ideas. A wide array of ideas in military education is being shared online, but this practice is not always replicated between institutions. Enhanced sharing of best-practice curricula by outstanding academic personnel, new learning approaches, and new military theories must be a cornerstone of the future approach to military alliances and partnerships.

Future strategic engagement must extend beyond the sharing and exchange of activities of like-minded military educational institutions. Military education must partner with the organizations undertaking technical research, such as DARPA in the United States or the Australian Defence Science and Technology Group. Engagement must extend well beyond military research and development entities. It must also embrace civilian industry if Western military institutions are to remain at the forefront of technological understanding.

The Chinese have adopted this approach. Using a strategy of civil-military fusion, the Chinese have implemented a whole-of-nation approach to innovation to create and leverage synergies between defense and commercial developments. The 2017 Chinese *Thirteenth Five-Year Science and Technology Military-Civil Fusion Development Special Plan* describes the requirement to "firmly establish and implement innovation, coordination, open, shared development concept, led by [an] innovation-driven development strategy and military-civilian integration."[47] As Elsa Kania has written, "In certain respects, military-civil fusion can be described as China's attempt to imitate and replicate certain strengths from a U.S. model, but reflected through a glass darkly and implemented as a state-driven strategy. Whereas China's initial aspiration was to progress towards an initial integration or closer combination of the defense and civilian economies, this new focus on fusion implies a deeper melding than had been previously achieved."[48]

Western nations should embrace a similar approach if they hope to compete with Chinese innovation and rapid adaptation and absorption of civilian and military technologies. This approach will facilitate a deeper appreciation of when and how curriculum adaptation is to take place and of the levels of technological literacy required at various stages of the future joint officers' careers.

Engagement with civilian universities is critical. Within these institutions reside hundreds of years of learning across the humanities and sciences. With the exception of some of the larger U.S., Chinese, and Russian military education institutions, most military educational institutions cannot hope to replicate such a knowledge resource. These civilian universities can provide intellectual rigor to further hone the intellects of military personnel at the undergraduate and postgraduate levels. Tertiary education is an excellent forum for developing critical thinking and complex problem-solving skills. They might also provide viewpoints on national security that differ from officially sanctioned policy, forcing military students to more carefully analyze the shibboleths of contemporary national security policy.

## Creativity, Futures, and Science Fiction

Thinking about future challenges and future military operations is a core responsibility of military institutions and their leaders. Indeed, there are many fine institutions in Britain, Singapore, Russia, and the United States that dedicate themselves to this mission.

Largely, these institutions have invested in understanding scenarios and technologies for future conflict and strategic competition. These are important undertakings and provide governments and military institutions with a useful foundational knowledge when developing or adapting national security strategies to synch with their military budgets.

Efforts to develop a range of plausible futures for military institutions must widen the focus on people—their training and education as well as the organizations and ideas they will use. These areas are covered to a degree in contemporary reports, but not as an area of focus or in the same detail as emerging technologies. For example, the 2018 *Global Strategic Trends* document from the British Defence Concepts and Doctrine Centre devoted only 24 pages to human development issues in a 270-page document.[49] The U.S. National Intelligence Council 2017 equivalent, *Global Trends: Paradox of Progress*, does a little better; it mentions or discusses the term "education" on 49 of its 226 pages.[50]

Professional military education must also be designed to meet the needs of military leaders and planners for the future environments that they are likely to face. To remain relevant, the curriculum of military institutions must be informed by an institutional viewpoint with respect to the type of future environments in which its people will operate. Military educational institutions must therefore form closer and more substantial linkages with organizations—in the military and beyond—that undertake futures work. There should be a transparent and logical pathway from informed views of the future and the type of intellectual development received by the joint military leader.

Like civilian education, most military education focuses on providing students with a set of predetermined competencies, such as basic military skills, team-based tactical activities, writing computer code, or studying the use of language. Anticipating change in the security environment will permit educational institutions to assess new skills their people will need in the future. This will allow military institutions to "future-proof" their curricula and ensure their students are optimally prepared for a more uncertain and challenging security environment.

Additionally, the link between futures on the one hand and education and training on the other allows students to gain a baseline level of skills in futures studies. Understanding this link will assist in developing a strategic mindset. It will also ground the development of an individual's capacity to draw conclusions on future threats and opportunities to supplement what is often a limited institutional futures capacity.

But we might also take this approach further. Perhaps we can peer into the more distant future and nurture the creativity of military personnel through the use of science fiction. Over the last decade, multiple military institutions have explored the study of science fiction in their thinking about future conflict. In the wake of the September 11 attacks on New York and Washington, DC, the U.S. military very quietly hired Hollywood directors and writers to "think differently about future threats."[51] The Canadian Defence Force was one of the first to do so more openly. In 2005 as part of its examination of future military challenges, it commissioned science fiction author Karl Schroeder to write the novella *Crisis in Zefra* in order to "illustrate emerging concepts and technologies that could become part of Canada's army in the future."[52]

In 2013 the Atlantic Council commissioned a project led by author August Cole to examine future conflict through the lens of science fiction. The project resulted in a series of short stories in a 2015 anthology called *War Stories from the Future*. The introduction by former Chairman of the U.S. Joint Chiefs Gen. Martin Dempsey notes that "at their best, science fiction stories explore the art of the possible, illuminate problems we might otherwise overlook, and entertain us at the same time. It sparks the imagination, engenders flexible thinking, and invites us to explore challenges and opportunities we might otherwise overlook."[53]

Since then, other institutions have embraced this approach. The U.S. Marine Corps employed science fiction futures in a 2016 project to review the future security environment. In 2018 the Australian Defence College formed the Perry Group, an elective designed to use science fiction to think about future force structure and warfighting concepts.[54] In 2019 the French Defence Innovation Agency began hiring science fiction writers to "form a red team that will come up with scenarios of disruption."[55] Finally, authors Peter W. Singer and August Cole have stimulated debate, and the imaginations of military personnel around the world, with their near-future military science fiction books *Ghost Fleet* and *Burn In.*[56]

Science fiction is useful in preparing our military leaders for several reasons. First, it shifts the paradigm and helps them to think beyond the day-to-day aspects of training, administration, and unit management. Second, it broadens their intellectual scope by exploring potential "bad futures" for military institutions. Science fiction also broadens the mind of future leaders and forces them to read something other than current events and military history. It does not replace these topics but complements them. Finally, it serves to inspire people and provide them with different or more optimistic visions of the future. Therefore, science fiction offers military organizations an additional tool for nurturing creativity and imagination in an otherwise conservative institution. It must be part of the military curriculum and the exploration of future concepts for all Western military institutions.[57]

## Educational Institutions as Think Tanks

Military organizations should not see their professional educational institutions as pure learning platforms for individuals. Military organizations (generally) select high-performing and high-potential military (and civilian) personnel to attend staff colleges and war colleges. Given the range of talent and the diversity within this large network of officers who have access to high-quality academic advisors, these student bodies have an expanding role to play in thinking through institutional responses to the strategic challenges that are disrupting national security establishments. We should therefore see military educational institutions as think tanks and idea incubators for the wider military institution.

A proportion of formal education should be focused on solving institutional problems. This will exploit the talent concentrated in educational institutions, while supplementing other dedicated institutional capacities to undertake force modernization. The output of such work could potentially comprise such things as student papers on service and joint future challenges, wargames and working groups on specific problems, or perhaps the development of more detailed reports by groups of students. Evidence from the history of the twentieth century shows that this is a useful approach. Perhaps the best documented example is that of the U.S. Naval War College in the years leading up World War II.

Fleet Admiral Chester Nimitz stated in a speech to the U.S. Naval War College in 1960 that "rigorous and repeated Naval War College wargames had ensured that nothing that happened during the war was a surprise . . . except the kamikaze attacks."[58] The Naval War College had been wargaming for decades by the time it became involved in examining the many different options and force postures for the U.S. Navy in the Pacific in the 1930s. It played a multitude of games, some with the Navy postured well forward in the Western Pacific, and others where the Navy took a more cautionary stance behind the line of Alaska-Oahu-Panama.[59]

The war college was also one of four pre–World War II centers for the study of amphibious operations for the U.S. Navy, although the Navy at that time did not see amphibious operations as critical.[60] But by 1930, the war college undertook annual wargames, which Allan Millett has called a "grand production that included Navy and Marine faculty and student officers from Newport and Quantico."[61] Coupled with a dedicated amphibious assault force—the U.S. Marine Corps—and an appreciation that island base seizure was a likely future mission, these wargames informed institutional learning and prepared a generation of leaders for future conflict.

The U.S. Naval War College in the interwar years provides a shining exemplar of educational institutions as think tanks for the military. There are also good examples of this occurring in the contemporary environment. The Naval War College undertakes wargaming to support U.S. Navy concept development.[62] The U.S. Army War College runs a very useful blog on contemporary military issues.[63] However, these examples represent outliers that should be employed more broadly across the entirety of educational institutions as part of the development of military leaders.

While the educational and training establishments in military organizations cannot carry the entire weight of analyzing future warfare, they have a vital contribution to make. With this contribution, they are ensuring that their parent military institution is better prepared for future conflict. They are also guaranteeing that their military leaders will have a more robust intellectual armor for the challenges—both known and unknown—that they will face in the coming decades.

## Adaptive Training and Education

One of the most serious obstacles to change in military organizations is the fact that they are human institutions. Many elements possess a bureaucratic mindset. This bureaucratic approach is not all bad; bureaucracies are excellent at imposing order and processes on an environment that is inherently ambiguous, disordered, and constantly changing. It is this constant change that poses the biggest challenge to institutions. As MacGregor Knox writes, "Bureaucracies are neatly *zweckrational*: swift and precise—in theory and surprisingly often in practice—in executing orders. They are happiest with established wisdom and incremental change."[64]

However, military organizations and their bureaucracies must keep up with the rapid pace of change. Military leaders, now and in the future, owe more to their people in response to this changing environment than established wisdom and incremental change. Retaining such an approach would represent the high point of slovenly thinking—at least until we lose the next war.

To be optimally prepared for an era of competition and conflict, and to retain relevance and stay at the forefront of best practice, the system for developing military leaders must possess mechanisms to adapt. While international conferences and seminars are a useful start, more research on learning and academic engagement in the military environment is required. This research might be conducted by individual institutions, but a more productive and effective approach might be for military institutions and academia, or even the military and academic institutions of multiple nations, to conduct it collaboratively. Noting the trends of rapid change, we must also better understand how we can effectively retrain our workforce more often.

Military training and education institutions must possess formal mechanisms to identify the need for change, to make informed decisions about it, and to enact those changes in a timely and efficient manner. They will need to be endlessly creative and constantly adaptive in the twenty-first century.

### *New-Era Learning*

Distinct from the organizational changes examined in the previous section are the educational initiatives that will be necessary in the coming decades. Military educational methods reflect a heritage approach—training-focused with long gaps between episodic educational interventions. Given the rapid pace of technological change and the evolution of the demographic profiles of the societies from which military personnel will be drawn, military institutions must also change their approach to professional military learning.

Education provides individuals with the enabling skills, knowledge, and attributes necessary to undertake the full range of military tasks and includes activities for developing communication and thinking skills. Education develops thinking processes that allow individuals to connect their training with the situations in which they find themselves and to apply the best course of action to deal with those situations. The broadening of individuals' horizons occurs through the education process, and this permits them to absorb their training more rapidly (a key twenty-first-century requirement) and with enhanced understanding. Finally, it is through education that military institutions will develop the individuals and leaders who can think clearly, apply knowledge, solve problems under the stresses of uncertain or ambiguous conditions, and communicate these solutions.[65]

Military organizations must ensure that the education available to their people is appropriate to their learning needs (the individual intellectual edge) and those of the military institution to which they belong (the institutional intellectual edge). The key elements of an evolved approach to military learning incorporate continuous learning, enhanced technological literacy, guided self-development, more rapid learning and relearning approaches, and elite programs.

## Continuous Learning

Rapid technological change complicates the ability to anticipate which job skills will prosper and which will become redundant. The nature of work continues to evolve, with the creation of new industries and job families occurring more quickly and older job categories becoming obsolete very rapidly.

In the contemporary era, rapid technological advances are making old skills obsolete and requiring new skills almost overnight.[66] This demands a shift from older episodic models of training and education to a more continuous model. Indeed, continuous learning may be one of the most important adaptations that military institutions could make in an era of hastening technological change.

In their examination of military innovation during the interwar period, Williamson Murray and Allan Millett found that military leaders were better able to lead and invest in innovative new ideas and technologies when they had undertaken continuous learning throughout their careers. They noted that "professional military education must remain a central concern throughout the entire career of an officer." By doing so, military institutions might "foster a military career where those promoted to the highest ranks possess the imagination and intellectual framework to support innovation."[67] Therefore, for the optimal development of the future military officer, military institutions must avoid long temporal gaps in formal professional development activities.

Military institutions require a continuum that possesses functional descriptors of what their people must be capable of at various stages in their professional journey, and a description of the curriculum that underpins the model. Nevertheless, that curriculum should be not an industrial-age production line but rather a nucleus around which we can build individually tailored intellectual development. This would also provide intellectual coherence to a more integrated approach to warfighting and achieving national security end states.

The desired outcome is a system that inculcates in military personnel a culture of continuous learning and an insatiable curiosity about the profession of arms.[68] For the optimal development of our military leaders, their desire to learn must be ignited upon entry to the military and continually nurtured, avoiding long temporal gaps in formal professional development. Experience does not adequately fill intellectual gaps. It is also less effective without the theoretical foundations provided by military education and training.

## Enhanced Technological Literacy

By building technological literacy in our military leaders, they will continue to grow a deeper appreciation of the application and impact of disruptive technologies on the wider context of joint integrated conflict and competition. There is a compelling need to ensure that military institutions continually enhance the technological literacy of their military leaders.

Building technological literacy focuses on military capabilities and new technologies, as well as on how institutions can absorb these technologies into their structures and doctrine to form new or improved capabilities.[69] This area of study also challenges our personnel to understand emerging technologies and to appreciate the subsequent threats and opportunities. This is a responsibility of military institutions but is set against the context of the rapid technological development in the surrounding society. A 2018 U.S. government strategy for science, technology, engineering, and mathematics (STEM) describes individual success in the twenty-first-century economy being reliant on STEM literacy. In a world of rapidly changing technology, people will need to encapsulate the capacity to effectively leverage digital capabilities and STEM skills such as evidence-based reasoning just to function as an informed citizen or worker.[70]

The requirement for enhanced STEM qualifications in the broader national workforce has been analyzed over the past decade. A range of studies have highlighted the need for an increased proportion of an organization to have acquired academic technical mastery and for improved technical skills in the broader workforce. The military is no different; it too will need to expand the uptake of specialized technical proficiencies in its personnel. As military equipment increases in technical complexity, its operation will require more technically qualified personnel and enhanced technical competencies within the entire workforce.

## Guided Self-Development

In 1982, a young U.S. Marine Corps officer, Paul Van Riper, wrote on the requirement for self-study to complement formal training and educational programs for officers. He noted that "the responsibility for professional development between periods in formal programs rests with the individual officer. This is inherent in the nature of the military officer's calling. It is inherent because the body of knowledge which constitutes the art and science of war is not only broad and deep but is also dynamic."[71]

There are three important ideas in Van Riper's essay. First, formal education in military institutions cannot cover all the needs in the intellectual development of our military leaders. The deduction from this assertion is that military institutions must complement formal education with self-initiated study regimes. This is not a new idea—when Gerard von Scharnhorst established the first command and staff college in Berlin (formal education), he also established the Military Society (informal education).[72] Informal education, facilitated by connectivity, has been elevated to a whole new level in the Internet age. It is explored later in this chapter.

The second element to draw from Van Riper's work is that self-study is likely to be most effective if it complements formal educational programs. The implication of this is that military organizations should provide curated online resources that military personnel can download from their institution, using the Internet or other resources, to supplement their professional development. This has occurred in many military institutions over the past decade across the globe, particularly in the United States and Australia.

The final implication from Van Riper's essay is that professional development material provided by curated hubs can be changed quickly. This is an important characteristic. Even the most adaptive military institution can only move at the pace of its bureaucracy. Therefore, less formal learning activities may provide a resource with a shorter adaptation cycle than the curriculum in military schools and academies. As a consequence, these curated resources for self-study are proving to be an increasingly important element of any military education system in nurturing an adaptive approach to the intellectual development of future leaders.

## Rapid Learning and Relearning

A range of studies over the past decade from international institutions such as the European Union and the World Bank have reinforced the speed at which workforces within nations may have to be retrained or reeducated. As one European Union report has described, "The key insight for the future is that all citizens will need to continuously update and enhance their skills throughout their lives, from the cradle to the grave."[73]

More recently, Yuval Harari has forecast that the "automation revolution will not consist of a single watershed event, after which the job market will settle into some new equilibrium. Rather, it will be a cascade of ever bigger disruptions. Old jobs will disappear, and new jobs will emerge, but the new jobs will also rapidly change and vanish. People will need to retrain and reinvent themselves not just once, but many times. In the twenty-first century they will need to establish massive re-education systems for adults."[74]

To develop personnel that can function and excel in this environment, future military institutions will need a system built around rapid retraining of their personnel as new technology and strategic circumstances arise. This is not the case in most contemporary military training and education systems.

## Elite Programs

During the past two decades, institutional education programs have had to balance their focus between two competing priorities: honing excellence in top-tier military officers, and catering to the developmental needs of a larger swath of very good officers who will fill the staff officer and midlevel managerial and leadership positions throughout the wider military organization.

In recent history, many talented military officers have been able to undertake yearlong studies at elite civilian universities. Nevertheless, elite study is typically outside normal staff college and war college curriculum. Professional military education (PME) has focused on a mass approach and has avoided elitism in mainstream courses such as those at command and staff colleges.

A contemporary approach to development of an elite form of military thinking is the advanced warfare courses in several nations. Among the first was the U.S. Army School of Advanced Military Studies (SAMS) in Fort Leavenworth. The school "provided a second year of graduate-level education in the science and art of war to a select group of majors, who the army would then seed into key planning positions. Its first class graduated in June 1984 and soon showed their value to senior Army leaders, who vied to get graduates on their staffs. While maybe not quite the 'Jedi Knights' they were touted to be, SAMS graduates made significant contributions to operational level planning."[75]

Institutions such as the U.S. Marine Corps and the U.S. Air Force have replicated this approach, as have institutions in Australia. More recently, the Johns Hopkins University established the "Art of War" program.[76] While the conduct of this elite education will continue to evolve and adapt to changes in the strategic environment, it is an indication of the value that some military institutions place on advanced intellectual development. This culture of demanding intellectual excellence—professional mastery—is an important element of building an institutional intellectual edge.

If military institutions are to realize the full potential of their military leaders in the twenty-first century, especially those with those most potential for strategic-level appointments, they have two options. First, they can establish elite courses such as SAMS, targeting more senior military and civilian personnel who demonstrate exceptional potential for strategic planning and leadership appointments. If this is not palatable, they may need to refocus their curricula on the most talented of their students in the mainstream staff college and war college courses.

However, well-resourced educational initiatives will not be sufficient to prepare military leaders for the challenges they will face. There is one more element of the trinity of initiatives: Military institutions must leverage the full range of new and disruptive technologies to underpin institutional learning and to personalize lifelong learning.

### *Technological Initiatives*

Building an effective system for developing military personnel in the twenty-first century will demand investment in technological initiatives ranging from broadening access to learning to enhancing connectivity between communities of interest. There will also be a need for more sophisticated undertakings that expand our understanding of human learning through to potential augmentation of human cognition.

### Improved Access to Joint Education and Training

Much contemporary joint education and individual training is delivered in a residential setting. While this model facilitates good learning outcomes, it results in only a small percentage of military personnel having access to these opportunities. The rollout of a joint continuum will provide the framework for development of online and self-study options for military personnel. This will be particularly beneficial where the continuum informs the personnel that there will be large temporal gaps in current joint learning opportunities, or where unit collective training imperatives affect whether personnel can be released for attendance of residential courses.

A consequence of the recent COVID-19 pandemic is that nearly every PME institution across the United States, Canada, the United Kingdom, Australia, and other nations transitioned to fully remote learning. Some military institutions with extant learning networks and online resources were well prepared for this; others were less prepared. Notwithstanding the fact that some military institutions were far ahead of others with respect to remote learning, most will freely admit that they learned invaluable lessons from the rapid transition from residential learning to remote platforms. These lessons, which have been shared among different nations, have included better designed and more robust learning networks, better accessibility for learners from their home locations, better preparation for instructors and academics who are teaching over these networks, and more secure teleconferencing capabilities that are designed for large and small group interaction.[77]

Lessons from the COVID-19 era for professional military education can inform an evolved approach to delivery of learning outcomes. Military institutions must continue to break down geographic, technical, and cultural barriers to create a truly connected force where education is continuous and self-sustaining. The system we design to develop our military leaders must be readily accessible not only to them, but to all military personnel, regardless of role, rank, or location. It should be applied judiciously across a joint force to build intellectual capacity at every rank, to include regular and reserve military personnel as well as civilian personnel.

The future infrastructure that will underpin this systemic approach to professional military education must allow military personnel to learn residentially, independently, or collaboratively, in a secure environment regardless of their physical location.[78] This expanded access must include personalized online education as well as unit-based PME and professional development activities.

## Engagement in the Global PME Ecosystem

Future systems to develop military people will exist within a global ecosystem of formal and informal mechanisms that hone the intellectual capacity of military personnel. The recent growth of noninstitutional media such as blogs, podcasts, and social media–enabled networking is likely to continue to expand in the coming decades. Military organizations can exploit the resulting informal network of military and national security professionals more effectively. This would allow military organizations to engage learners in different ways and provide a wider range of learning materials, even if these resources do not always neatly fit within the formal military educational continuum.

Formal networks among military institutions are an element of an effective global PME ecosystem. While annual conferences, such as those hosted by the NATO Defense College, provide useful networking opportunities, they do not fully exploit the capacity to share ideas and new methodologies among those people who will oversee the intellectual development in military officers. A more expansive range of exchanges between military officers and academics from diverse military institutions should be considered. After due consideration, many of these initiatives could be implemented to facilitate a wider sharing of ideas and learning with respect to the divergent approaches available to support the intellectual development of national military institutions.

## Innovation in Engagement and Delivery

The rapid pace of change in technology has impacted societies around the world. It is also disrupting long-standing approaches to training and education within military institutions. Technology has enabled a more connected approach to learning—particularly the use of blogs, curated self-study sites, online learning and massive open online courses, and video conferencing.

These advances in technology-enabled learning have provided new capabilities for teaching and learning. New learning approaches using online technologies are available for students who may have been excluded from existing models of military education. As digital learning environments become more sophisticated, competency in educational technology must be a core skillset for teachers in military institutions. They must become digitally savvy in designing learning activities and courses that leverage these new technologies.

For students, these technologies present an expanded range of opportunities to learn through their own discovery and through collaborations.[79] As the explosion in the number of military-themed blogs and self-study sites such as *The Forge* and *The Cove* attest, today's junior leaders have a significant appetite for self-discovery and learning to complement more formal military educational experiences.[80] Military organizations need to exploit this appetite to learn by self-investigation and develop tangible mechanisms to incorporate these technologies into their institutional designs to produce military leaders.

Digital-age technologies offer more advanced approaches to learning, but AI also has the potential to significantly change the way militaries educate their personnel. AI might underpin an expanded range of activities to support learning, such as intelligent tutoring systems for individuals,[81] or provide a better-informed learning partner that accompanies and assists the learning of future leaders throughout their career.[82] We are only at the beginning of realizing the benefits of AI in training and education; however, in the future, it could possibly assist human cognition.

## Augmenting Human Cognition

Augmenting our biological sources of intellectual capacity with artificial intelligence may offer enhanced advantage for nations in the twenty-first century. While this potential might still seem like science fiction, breakthroughs in information technology and synthetic biology offer the prospect for augmenting human cognitive functions in the very near future. Fundamentally, augmented cognition seeks to circumvent human limitations by building environments where it is easier for humans to encode, store, and retrieve information presented to them. While new breakthroughs in artificial intelligence, data science, and biotechnology are providing a foundation for augmented cognition, it is the pace of change that is the true driver.

The field of augmented cognition incorporates topics such as adaptive learning systems, brain-computer interfaces, crowd-augmented cognition, cognitive load and performance, and neurotechnology. All these areas are

still relatively immature. However, one promising area for early examination is the concept of AI extenders. Proposed by Jose Hernandez-Orallo and
Karina Vold, AI extenders are an approach where artificial intelligence provides for the extension of human cognition. It is important to note that this
is truly an extension of human cognition with artificial intelligence; there is
no autonomy for the artificial intelligence involved.[83]

Artificial intelligence and the extension of human cognition may be
both desirable and possible for multiple functions in a military institution.
Building on the work of Hernandez-Orallo and Vold, I propose that there
are five cognitive functions that should be prioritized for augmentation or
extension. These would comprise basic cognitive functions that our people
and leaders apply routinely: enhanced memory, attention and search, comprehension and expression, planning and executing activities, and metacog-
nition. For more rapid and better-quality planning and decision-making at
both the tactical and strategic levels, using AI to assist human decision-making
in these areas would be desirable. It would not constitute an unethical use of
artificial intelligence; humans would still be the decision-makers, including
where the death of or injury to human beings was involved.[84]

Another possible source of improving human cognition is through neurotechnology. The potential for neurotechnology to improve human cognitive
functions is currently being explored by DARPA. In the Next-Generation
Nonsurgical Neurotechnology program, scientists are seeking to build a
neural interface that can read and write to multiple points in the brain at
once. These technological developments hold the promise for significantly
enhanced human-computer interaction in the future.[85] Combined with
advances in information technology, such as machine learning, bid data, and
artificial intelligence, this closer connectivity between humans and computers to improve cognition may significantly change how military institutions
are able to develop strategies and warfighting concepts.

Multiple obstacles to effective augmented cognition remain.[86] These are not
just technical obstacles; the ethical implications of such augmentation must be
fully explored. Despite this, a range of different approaches to augmenting—
not replacing—human cognition is likely to become available in the coming
years. The impact of these technologies on learning and developing military

leaders is worthy of exploration in the short term and may well be an area of collaboration across Western military institutions. Designing such technologies into the education, training, and development of future military people should commence soon if military institutions are to remain at the forefront of their exploitation.

## MILITARY SUCCESS THROUGH PEOPLE
## IN THE TWENTY-FIRST CENTURY

Military institutions in the West all face a similar spectrum of challenges. In the coming decades, the challenges of geopolitics, the future of work, disruptive technology, and climate change will drive how military institutions prepare their people to work effectively in this environment. The approach proposed in this chapter aims to stimulate institutional reconsideration of training, education, and professional development. We need investment in improving the intellectual development of military leaders so they are fit for the twenty-first-century security environment.

With appropriate focus, investment, and senior leader advocacy, the development of more effective military personnel (and their leaders) is possible and highly desirable. This new model military training and education system must operate by a strategic design to exploit history, new technologies, military and organization theory, and enhanced networking. Through the range of strategic, learning, and technological initiatives described in these pages, military institutions can provide the optimum investment in their people for future war and competition.

This chapter has aimed to respond to this challenge and to tell the story of how military institutions in the digital age can succeed—in peace and war—if they are prepared to rededicate themselves and their people to building an enhanced intellectual edge. With individual and institutional manifestations, this is a central tenet of military capability in an era of geopolitical competition and an accelerating pace of technological disruption.

Military institutions must construct this intellectual edge around the development of hyper-professional and elite military institutions, from the deployed force, to its education and training systems, to its strategic planning institutions. People in this system must be able to contextualize, plan, decide, act, and adapt faster and more successfully than adversaries.[87] The military must celebrate its elite thinkers, just as we might celebrate and value world-class artists or sportspeople. Institutional incentives must be broadened to encourage elite thinking. There are ample illustrations from our collective military history that prove what happens when military organizations fail in this endeavor.

The development of an intellectual edge is a cornerstone of military capability, regardless of national or institutional cultures. Given the profound changes in the strategic and technological environments, building and sustaining this edge for future competition and conflict will require changes in the delivery of learning by military institutions and a change in how they value this learning. This means that military institutions must do better to incentivize and nurture intellectual curiosity and celebrate their military intellectual elite. Conversely, those individuals who show a sustained lack of curiosity and disinterest in developing their intellectual capacity should not be promoted. This has educational as well as career management implications, some of which will clash with long-established approaches and cultures around the development of military officers.

Military organizations across the globe find themselves with a range of new circumstances affecting how their operations are conceptualized and executed and how they develop their people. In his book *The Big Stick*, Eliot Cohen quotes Abraham Lincoln, stating that "as our case is new, so we must think anew and act anew."[88] In developing an intellectual edge in our future leaders in these new circumstances, and then honing this edge so that it becomes an intuitional culture, military forces must also think anew and act anew.[89]

# CONCLUSION

## *War Is Transforming (Again)*

The assembled group of generals, their staffs, minor nobles, and foreign officers, along with the king of Prussia, Otto von Bismarck, and Prussian army chief of staff General Helmuth von Moltke, stood on the hillside west of the French town of Sedan. They had gathered to view the 250,000 soldiers of their Prussian Third and Fourth Armies in action against the French forces of Emperor Napoleon III.[1]

Within a day, the Prussians had prevailed. By 2 September 1870 the French and Prussian guns had fallen silent after two days of battle. The Prussian force had encircled the 130,000-strong Army of Chalons. With all attempts to break out of the encirclement ending in failure, the French army had capitulated.

The Prussians captured more than 100,000 French soldiers and 558 guns. Napoleon III capitulated to the Prussians and was dispatched to comfortable captivity near Kassel. With all major French forces either under siege or defeated, the Prussians were free to lay siege to Paris, which they did from late September 1870 until the French negotiated an armistice in January 1871.

The French defeat was a result of the reforms undertaken by the Prussian military in the nineteenth century, including their adoption of new technologies. After their defeat by Napoleon in 1805–6, the Prussians had formed a military reorganization commission under Gerard von Scharnhorst. He undertook to "raise and inspire the spirit of the army, to bring the army and the nation into a more intimate union."[2] Scharnhorst and others had modernized recruitment, training, and officer education. General von Moltke had implemented other reforms in organization, mobilization, and tactics after he became Prussian army chief of staff in 1858.[3]

Michael Howard noted that the Prussians prevailed in the war because they were able to better exploit the changes both in technology and within their societies more broadly throughout the nineteenth century: "The social and economic developments of the past fifty years had brought about a military as well as an industrial revolution. The Prussians had kept abreast of it, and France had not. Therein lay the basic cause of her defeat."[4]

Just as the Prussians kept abreast of change in their time, so too must contemporary military institutions remain at the leading edge of their profession. They must constantly hone their readiness, deployability, and capability to be able to undertake a broad spectrum of operations. Military institutions cannot mark time and remain overly focused on being competent in current operations.

To increase the probability of future success, military organizations must seek out potential threats and then iteratively design and build the forces to meet those challenges. The effectiveness of people, ideas, and institutions will in large part determine whether military organizations can successfully adapt to meet the demands of the future security environment.

———

Changes brought about by the first Industrial Revolution had far-reaching impacts on the technology of war. They inspired new visions of war from theorists such as Antoine de Jomini and Carl von Clausewitz. New professional institutions were established to develop military personnel. A series of military social revolutions reshaped the relationships between military organizations and the nations they served. Nations were forced to confront the concept of total mobilization for war.

The second and third Industrial Revolutions saw more changes in technology, geopolitics, and society. These had follow-on effects in the preparation for, and the theories and conduct of, war. New institutions such as air forces were born. The conduct of total war was realized and, ultimately, weapons such as atomic bombs and biological agents were developed that could extinguish all life on our planet. The impacts of these developments were multigenerational and continue to be felt in the contemporary era. Theories of war at sea (such as those by Alfred Thayer Mahan) and in the air (by Giulio Douhet and others) remain powerful and influential in military institutions. In the case of new warship designs, trains, planes, and communications, the impacts have continued over a century.

The nascent fourth Industrial Revolution will result in more change for military institutions. Ongoing developments in technology will affect societies and change the character of war and strategy. These changes will endure beyond some of the short-term fads (think effects-based operations and "shock and awe") of the past two decades. This book has explored the shape of these changes. In it, I have proposed ideas for developing military institutions so they can operate effectively in contemporary and future security environments. It has explored solutions that will endure or, at a minimum, prove sufficiently adaptable to meet the demands of a continually evolving world.

Three important themes have permeated this exploration. The first theme is that war is an enduring part of human existence, but it evolves with developments in society, technology, and politics. We are currently in a period where technological innovation and change in society are surging. This is having an impact on how military institutions are organized and how they think about their activities. Similar to previous industrial revolutions, disruption caused by technology, geopolitics, and society is having an effect on human competition and conflict. The impact is now discernible. But there is much more for us to learn so we can understand the full range of implications of these disruptions.

A second theme is that changes in society and geopolitics are driving change in how military institutions must think and organize themselves. Qiao and Wang described this in *Unrestricted Warfare*:

The great fusion of technologies is impelling the domains of politics, economics, the military, culture, diplomacy, and religion to overlap each other. The connection points are ready, and the trend towards the merging of the various domains is very clear. Warfare is now escaping from the boundaries of bloody massacre, and exhibiting a trend towards low casualties, or even none at all, and yet high intensity. This is information warfare, financial warfare, trade warfare, and other entirely new forms of war, new areas opened up in the domain of warfare. There is now no domain which warfare cannot use, and there is almost no domain which does not have warfare's offensive pattern.[5]

War and competition now pervade nearly every aspect of human endeavor. There is almost no facet of human life that cannot be influenced by those governments that free themselves of the ethical and legal obligations accepted by liberal democracies. War and competition are now more integrated, with many more military and nonmilitary activities happening concurrently. We must reconceive how we think about military power. In an era where humans and machines will work together as more collaborative teams, new ideas of fighting, influencing, and competing will be required at every level of military endeavor. Nations and their military institutions will also require long-term agile strategies complemented by an intrinsic capacity to rapidly adapt at all levels to foreseen and surprising changes in this new environment.[6]

The final theme is that people are at the heart of war and competition, and they are the core of every challenge and every opportunity in this new and disruptive environment. It is people who lead military institutions and their units, and it is people who develop the various strategies that guide the long-term development of new forces and the execution of military campaigns. It is also people who surprise other people. Surprise—at every level of societal endeavor, including conflict—is more likely in the future. The 2020 COVID-19 pandemic has acted as an accelerant, speeding up (or making more obvious) changes in the global system.[7] We will need to change, and in some circumstances transform, how we train, educate, and lead our people for this twenty-first-century security environment.

Leaders—national, societal, and military—cannot be passive bystanders during this period of rapid change. There is much that nations and their military institutions might do now, and in the coming decades, to thrive in such a competitive environment. Military institutions must not be viewed as ends in themselves; they exist solely to meet the policy requirements of our elected democratic governments. But if they are to be effective and retain the capacity to meet the needs of our elected leaders, they must transform how they think about what they do, how they do it, and how they integrate their activities into the wider span of national power.

This book concludes with a summary of ideas—in the form of six propositions—that might assist the thinking of military institutions as we move forward through the challenges of the twenty-first century. I would not be so bold as to call this a road map for success. But it does provide a series of ideas that can help strategists, policy-makers, and military leaders think about how they can adapt to remain effective tools of national power.

## WAR AND COMPETITION WILL REMAIN HUMAN PREOCCUPATIONS

Since humans first walked the earth, they have been in competition with one another. Unless something fundamental changes in the make-up of *homo sapiens*, this is unlikely to change. Whether it is the myriad of individual behaviors and national responses we witnessed during the COVID-19 pandemic or the activities of various actors seeking to maintain or enhance their global influence, "fear, honor, interests" will remain the ingredients for human cooperation, competition, and conflict at both the individual and national levels.

We must therefore persevere with investing in thinking about national security and military strategies, the ideas for concepts and organizations, the best possible technology for military hardware, and the best training and education for military personnel. Military people, and especially their leaders, must have a deep understanding of the phenomenon of war, its many guises, its subordination to policy, and how its threat might be used to deter and influence competitors and adversaries.

At the same time, we must also ensure that the building and application of military power are always aligned with political requirements. The political system must continue to drive the purpose and aims of war. As Colin Gray has written, "Strategy should be thought of as glue that holds together the purposeful activities of state. It can be about many things, but primarily it must be about politics."[8] In *On War*, Clausewitz wrote that "the political object is the goal, war is the means of reaching it, and means can never be considered in isolation from their purpose."[9]

We must never lose sight of the fact that military power only exists in democratic systems to serve the elected government of the day. The imperative for this is only reinforced by the power that is derived from the legitimacy of military forces acting in accordance with the will of elected civilian governments and the values of the people they serve.

## FUTURE MILITARY POWER IS NOT JUST ABOUT THE MILITARY

The possession of a professional, lethal, ready military organization will remain central to states retaining their sovereignty and pursuing their interests. A capable and ready military institution will always be an important part of any nation's deterrent strategy and an essential aspect of their statecraft.

Nonetheless, not every national security problem has a military solution. Indeed, many will not. As Lawrence Freedman notes, "The realm of strategy is one of bargaining and persuasion, as well as threats and pressure, psychological as well as physical effects, and words as well as deeds."[10] Future military institutions must be designed within a more holistic national approach that encompasses security, competition, war, and the retention of national sovereignty as a national endeavor.

Potential adversaries have embraced strategies that focus on exploiting the seams in our societies and our military institutions. We must invest in a significantly improved integration of all elements—military and civilian, informational and commercial—to deny this attack vector to our adversaries. Deeper integration of the various national security organizations will underpin military activities as well as provide a national foundation for responding to other challenges such as pandemics and severe weather events. This may demand the rebalancing of resources across the national security enterprises of many nations.

Within military institutions, it may also necessitate a new approach to what we currently understand as "joint." We need to be a much more integrated force. This extends well beyond the joint approach that coordinates the efforts of air, land, and sea fighting forces. It extends into newer domains such space, cyber, and broader influence capabilities. But this integration must also encompass more enhanced interoperability between military forces designed for foreign activities and other government agencies that operate domestically.

A more connected and integrated method magnifies the individual capability each contributor brings to the entire national security effort. By unifying our efforts, we will close capability, institutional, and cultural gaps that our adversaries could seek to exploit. Military forces therefore must be designed, from first principles, to integrate. They must be able to adapt or morph depending on the circumstances they face and to operate effectively alongside other forces with different military capabilities, doctrines, and cultural backgrounds. This will require cultural empathy, deft balancing, and sound strategic decision-making. At the same time, military forces must remain interoperable with their allies and be capable of operating reciprocally with regional neighbors. In the end, however, a joint force needs to be joint in both its design and its practices. Without continuous individual and collective joint education and training, future military forces will not be able to capitalize on the benefits of an integrated design.

At the same time, the capacity to mobilize all aspects of national power is a key part of this integrated approach. Warfare is a whole-of-nation endeavor; we must be better at mobilizing the national capacity to both deter and pursue war. Western nations need better mobilization strategies that develop technologies and build large quantities of them. More importantly, they must also be able to mobilize a large proportion of the populace for different capabilities beyond military service. New forms of national service across a range of military and national security capabilities will also need to be examined and instituted. This will assist governments in responding more comprehensively (and cleverly) to the challenges of the new era of strategic competition, as well as to those unique and often unpredictable challenges posed by climate change and pandemics.

Contemporary technology has enabled around-the-clock confrontation, competition, and conflict, and this trend is likely to intensify further in the future. This means some of our traditional organizational approaches and decision-making models will no longer be applicable or effective. We will be challenged to optimize our structures and processes within and beyond military institutions in order to meet the demands of a rapidly changing pace of military and other national security activity.

## MILITARY INSTITUTIONS MUST WIN THE
## ADAPTATION BATTLE AT ALL LEVELS TO SUCCEED

In particular, our capacity to rapidly adapt ideas and institutions will be decisive. As we have explored throughout this book, we are experiencing a surge in technological and geopolitical change. While this may not last the entire century, the experience of previous industrial revolutions indicates that it is likely that we will have to deal with this rapidly changing environment for at least the next few decades.

One of the more significant implications of a rapidly changing environment is that surprise remains a potent and constant element. Whether it is the strategic surprise of breakthroughs in key technologies or the operational and tactical surprise that might be the outcome of military activities, it is an enduring element of human competition and warfare. It will not disappear. Therefore, an important aspect of generating military power in the twenty-first century will be improving the capacity of institutions and military units to adapt.

Adaptive capacity is required at every level but becomes more difficult as we rise through the higher levels of an institution. At the coalface of military operations—the tactical level—the imperative for soldiers, sailors, marines, and aviators to adapt is very clear. If they do not change based on the situations at hand, they are likely to fail. The manifestation of this failure is often death or disability. Therefore, the driver for effective adaptation at the tactical level is apparent and often requires little command intervention (with the exception of sharing lessons).

But as we rise through the higher levels of military units and institutions, the imperatives to adapt are less clear. In this book we have explored how "bureaucracy does its thing" and that process and incremental change are the true DNA of large institutions—military or civilian. Therefore, there is a need for reinforced drivers for institutional adaptation. These mechanisms include an improved environmental scanning capacity, the inculcation of an enhanced culture of risk awareness and risk-taking by leaders at higher levels, mechanisms to nurture, experiment with, and absorb new ideas, and fast-tracked research, development, and deployment of new technologies.

With a systemic approach to adaptation, institutions should have an improved capacity to develop new ideas. Opening up the capacity to develop, test, and absorb new ideas is a key outcome of an effective institutional approach to adaptation. It must result in a more rapid development of ideas, technologies, organizations, and doctrines that can be absorbed at the "speed of need," not the current "speed of committee."

An evolved approach to adaptation also assists in minimizing the chances of surprise and in responding when institutions or military forces are surprised. Surprise is a principle of war because it generates shock in the mind of commanders and other military personnel, and it leads to a level of paralysis. Those who generate surprise seek to exploit the situation during this following period of shock. But in an institution with a robust culture of adaptation, the period of shock might be reduced and the capacity of an adversary to exploit shock curtailed. And as we explored earlier, our understanding of adaptation can also help us to shape the adaptive processes of these adversaries.

So, adaptation is a core capability for military institutions. As Michael Howard has noted, success in warfare is less about being able to clearly predict or prepare for every eventuality than about being able to better learn and adapt based on how the interaction of combatants plays out. Howard describes this as a military organization's "capacity to get it right quickly when the moment arrives."[11]

## ETHICAL AND CREATIVE HUMAN-LED DECISION-MAKING BY HUMAN/MACHINE/ALGORITHMIC TEAMS WILL UNDERPIN FUTURE MILITARY EFFECTIVENESS

Many serious researchers in artificial intelligence have expressed doubt about whether true, human-like intelligence is possible in machines. As Luciano Floridi has written, "True AI is not logically impossible, but it is utterly implausible. We have no idea how we might begin to engineer it, not least because we have very little understanding of how our own brains and intelligence work."[12]

That said, algorithms are becoming more powerful. Coupled with increased processing power and data management, artificial intelligence can assist people by providing rapid and useful insights into complex, large-scale problems. Concurrently, autonomous systems are becoming more capable at a wider range of functions. They also are becoming cheaper to procure, in larger numbers, by state and nonstate actors. Because of this, military forces will induct and deploy tens of thousands of increasingly brilliant robots across the land, sea, air, and space domains. They will concurrently deploy algorithmic capabilities—artificial intelligence—into every military and national security endeavor. The decisions of human strategic leaders, policy-makers, and tactical commanders will be shaped and informed by AI.

The proliferation of robots and algorithms into human institutions will enable all kinds of new operating concepts and organizations. It will improve the performance of humans and teams, at various scales and in all domains. And it will potentially allow small nations to exert a military effect disproportionate to their size.[13]

But this will also have a disruptive effect on institutions that are by nature conservative and suspicious of unproven technologies and ideas. Existing career models, promotion pathways, command and control, and risk management must all be transformed. We can, and should, imagine a future where some military specialties, such as pilot and infantry, disappear largely from military institutions.

But just as we must be bold and creative and take risks with new technologies, we must not allow them to overwhelm our values and those things that make institutions human. War remains a human undertaking. Regardless of their sophistication, the tools of war support humans. This is an important point, especially in an era where many believe human decision-making might be overtaken by machines in the form of artificial intelligence. Even if this were possible, it is not a desirable goal in military institutions for a range of ethical and legal reasons.

Human decision-making, with all its flaws, must remain preeminent in any human-machine relationship. Machines, algorithms, and autonomous systems, as important as they will be in the future, do not have souls, values, or purpose. We would do well to remember this in our human-machine future military environment.

## PEOPLE ARE AT THE HEART OF ALL MILITARY ADVANTAGE IN THE TWENTY-FIRST CENTURY

Because of this, we must build a more robust intellectual edge in our people and institutions. While economics, diplomacy, information, and technology remain important aspects of national power, they are insufficient to successfully deal with the challenges that Western nations face in confronting well-resourced, strategically minded, techno-authoritarian regimes. This challenge is compounded by the fact that nations such as Russia and China have invested in developing new ways of thinking about warfare. As the two Chinese Colonels remind us, "While we are seeing a relative reduction in military violence, at the same time we are seeing an increase in political, economic, and technological violence. The new principles of war are no longer using armed force to compel the enemy to submit to one's will, but rather are using all means, including armed force or non–armed force, military and non-military, and lethal and non-lethal means to compel the enemy to accept one's interests."[14]

Because of the Russian interference in the 2016 U.S. elections and widespread Chinese influence operations, we have only begun to understand the full meaning of these words and their implications. Part of this is because the large military bureaucracies in the past two decades have been absorbed with the challenges of counterinsurgency operations in places such as Iraq and Afghanistan. Large-scale nation-state warfare has been a lesser concern in the academies and colleges in many military institutions. This has begun to change, but we must do more.

We will need to instill in our people, especially those who we might call our strategic and institutional leaders, a vastly better understanding of new and disruptive technologies. In some respects, most Western military institutions assume this away. We think that because our soldiers, sailors, marines, and aviators are trained on the expensive and complicated machinery of war, they possess this literacy. Unfortunately, this assumption is both false and dangerous.

New and disruptive technologies are changing quickly and are proliferating to many nations and nonstate actors. Technologies such as artificial intelligence, biotechnologies, new materials, and space-based capabilities not only change the character of warfare, they also intermingle and overlap in ways that are difficult to discern. Military institutions must provide their people with a significantly improved and ongoing education about the technological environment and how they can be integrated into new policies, new warfighting concepts, and new organizations.

Another aspect of generating the intellectual edge is trusting our people more and allowing them to have difficult conversations—behind closed doors and in public—about military and strategic concepts. While the primacy of civil decision-making in democracies is sacrosanct, we cannot use this as an excuse to shutter discussion about strategic options and new ideas. I have seen this too often. We must permit a more open debate with ideas and take risks by sometimes having uncomfortable conversations about institutions and ideas that are past their prime.

Military institutions must also continue to find new and improved ways to nurture in their people the desire to learn. An insatiable curiosity must be a first-order behavior in twenty-first-century leaders. It underpins creative, adaptive, and innovative leadership. Those who do not exhibit this trait, or who seek to suppress it, deserve to be thrown upon the rubbish heap of failed leaders who demonstrate a better capacity for process than achievement. Fortunately, there are many young leaders willing to assume appointments with more responsibility and lead in more innovative and energetic ways.

Military institutions will need to nurture a culture that embraces the intellectual pursuits of the military profession. Western military leaders rarely give speeches of the type that General Gerasimov gave during his tenure as chief of staff of the Russian military forces. This must change. Senior military leaders must incentivize the type of intellectual endeavor that will produce the ideas required for a new era of conflict, competition, and cooperation. The best thinkers within this renewed intellectual environment, the elite military thinkers, must be celebrated, recognized, and nurtured. Military institutions must offer incentives to encourage elite thinking and, potentially, promotion pathways and talent management to support this.[15]

## WE MUST UNDERSTAND HOW OTHERS
## SEE WAR AND COMPETITION

An important aspect of twenty-first-century military power will be to better understand and overcome cultural asymmetries. To do so, Western nations must better understand how other cultures, such as the Chinese, Russians, Indians, and others, view war and competition. Their appreciation of the nature, duration, and boundaries of war is often different to our own.

It took years for Western forces to gain some understanding of the cultures of the different peoples in Iraq and Afghanistan. In many respects, those of us who served there were hindered by a widespread lack of understanding about those we were supposed to be helping and those we were supposed to be stopping (the insurgents). David Kilcullen has written on the topic extensively. In his contribution to a 2019 book on military culture by Williamson Murray and Peter Mansoor, Kilcullen examines a vital asymmetry between Western nations and others. He notes that

> if Americans and their allies, with a narrowly defined cultural norm for [what] is and is not war, confront an adversary with a vastly broader cultural understanding of conflict, then two equally dangerous things can arise. First, we can be engaged in conflict with an adversary who considers themselves to be at war with us and yet not realize that fact. Second, we can be engaged in activities that seem innocuous or peaceful to us—routine deployments, support for peaceful democracy, expansion of military alliances—and our adversaries can perceive these as acts of war and respond accordingly.[16]

Misunderstanding and miscalculation led to many of the great conflagrations that engulfed humans in the past. While war may remain part of human affairs and perhaps even a necessity at times, we should do everything we can to educate our people about "the other." Developing the intellectual edge in our people and organizations must have as one of its pillars the cultural understanding of others. This means that we must study, to a degree we have never done before, the cultures and ideas of non-Western nations. This includes those we may fight as well as those we may fight alongside.

Overcoming our misunderstanding of others will be a fundamental undertaking for military leaders, national strategists, and military learning systems in the coming decades.

## CONCLUSION: WAR IS TRANSFORMING (AGAIN)

As Clausewitz noted, "War is more than a true chameleon that slightly adapts its characteristics to the given case. As a total phenomenon its dominant tendencies always make war a paradoxical trinity—composed of primordial violence, hatred, and enmity, which are to be regarded as a blind natural force; of the play of chance and probability within which the creative spirit is free to roam; and of its element of subordination, as an instrument of policy, which makes it subject to reason alone."[17] The form and function of war have transformed. It remains an extension of politics—"a true political instrument, a continuation of political intercourse, carried on with other means"[18]—but perhaps not in a form that Carl von Clausewitz would understand. In his era, war was almost an exclusively military activity, with limited contributions by diplomacy.

War will remain a truly national endeavor. It embraces all aspects of the life of a nation—economic, social, commercial, cultural, informational, diplomatic, and military. The "extension of politics" now has significantly broader and deeper implications, including the need for better integration of national security organizations in the conduct of their physical, and influence, activities. These impact not only statesmen and strategists, but also military and civilian leaders, the business community, academia, and the wider society. It has become the "total phenomenon" about which Clausewitz wrote.

War is also an integrated partnership of violence and influence. These two aspects are now inseparable and almost indivisible. Just as the geographic boundaries of war have fallen away—the home front and battlefront are fully connected—so too have the stovepipes that governments and military institutions have traditionally erected around violence and influence. Understanding this indivisibility of violence and influence allows political and strategic leaders to build more creative policies and strategies that will permit Western democratic systems to chart their path through the perils of the twenty-first century.

We must heed the lessons of past industrial revolutions. Each has in its own way transformed military operations, ideas, and organizations. The new and still unfolding fourth Industrial Revolution will do the same. But never before have we possessed the capacity to collect information and learn about this ongoing revolution. Never before have we also possessed the learning frameworks and digital analytical capacity to learn from these changes and apply solutions to them as the world changes around us.

In a 1973 lecture, Michael Howard used a wonderful metaphor about how military institutions must view themselves as adaptive and flexible organizations. He described how the military "must not see themselves, as they all too frequently do, as an old order defending civilized values against a revolutionary threat: people in an ocean liner peering uneasily out through the portholes at an increasingly stormy sea outside. They should see themselves as intelligent surf riders spotting the essential currents in a sea which is certainly disturbed and by no means friendly but on which, if they are skillful enough, they will survive."[19] The leaders of military institutions in the twenty-first century must be Howard's skillful surf riders so they might deter coercion and defend their nations in what will continue to be a turbulent geopolitical and technological era.

———

On 11 September 2001 I was a young officer attending the U.S. Marine Corps Command and Staff College at Quantico, Virginia. It was a beautiful day, but not for long. Along with most of the world, I watched in horror as nineteen murderous individuals flew hijacked aircraft into the Twin Towers and the Pentagon. And unbeknownst to us, the courageous passengers of a fourth flight turned on their hijackers and sacrificed themselves to prevent another potentially catastrophic attack, probably on Washington, D.C.

I watched this on television with my conference group classmates in a seminar room. On that terrible September morning, I saw on their faces the abject horror and helplessness of those subject to a surprise attack. I witnessed the disbelief that someone could hate them, and the nation they loved, so deeply that they would attack and murder thousands of innocent civilians at their places of work.

We knew then that our world had changed. We knew that someone was going to pay for this, and that some—or all—of us in that room would probably play some part in that. But none of us really anticipated the long wars that would follow, and our repeated deployments to places such as Afghanistan and Iraq. None of us fully appreciated just how profoundly things would change over the coming decade.

The United States and its allies have spent two decades absorbed in chasing terrorists and conducting nation-building operations. These operations cost trillions of dollars while killing and wounding tens of thousands of military personnel. Hundreds of thousands of civilians in the Middle East have also been killed and injured in the wars spawned by 9/11. Military operations over the past two decades, however, have also built a new understanding of the importance of nonkinetic operations and the power of influence that military organizations might exercise.

But others—nations such as China and Russia—were also watching the operations in Iraq and Afghanistan. They learned the weaknesses of Western conventional forces and the power of strategic communications and influence activities. Their observations in many ways reinforced their existing strategic cultures and inclination to deceive. And despite a recent shift of attention to the western Pacific, over the past two decades the nations in the West have avoided hard decisions about the true nature of future military power.

Authoritarian regimes whose ideologies suppress the potential of their people have also taken great comfort from the missteps or calamities of Western nations during the COVID-19 crisis. They have emerged from that crisis as a mighty challenge to democratic systems of government.

Hard decisions cannot be avoided. We must transform how we think about national security and how we design military institutions. How we conceive military power, within a joined-up approach to national power, will determine whether we retain a full measure of national sovereignty over the next century. We owe it to those we serve to transform our military institutions so we can retain the capacity to protect the freedoms we have fought for and earned.

Two days before 9/11, I spent the day with my family in Washington, D.C. It was a beautiful autumn day, and we spent a lazy morning wandering through museums on the National Mall. There are days when I wish we could return to that era. But life is lived forward. We must turn to the future and imagine how we can make that future the best that it might be for our people. It will require members of the profession of arms, and those in the wider national security community, to be more creative, more connected, and better able to build an intellectual edge over potential adversaries.

Clausewitz wrote that "every age has its own kind of war."[20] More recently, Chinese president Xi Jinping noted that "the wheels of history roll on; the tides of the times are vast and mighty. History looks kindly on those with resolve, with drive and ambition, and with plenty of guts; it won't wait for the hesitant, the apathetic, or those shy of a challenge."[21] We must better understand the types of competition and wars we will face in this new age of war. We need to develop the adaptive mindset to keep pace with threats and keep at bay the major threats to our societies. As war continues its transformation, we must not be hesitant or apathetic or shy from the challenge. The future of our democratic systems and the liberty of our peoples may depend on it.

# Epilogue

## *A Changing Nature of War?*

The central theme of this book has been the ongoing transformation in the character of war. However, before we conclude this exploration of future war and the role of people, ideas, and institutions, a final question—one that has gained some attention over the past decade—might be asked: Is the technological environment, which is rapidly evolving because of developments in artificial intelligence and biotechnology, changing the nature of war as well as its character?

Military professionals and academics differentiate between the nature and character of war. The nature of war refers to its essence—violence, driven by politics, and the fact that it is an interactive competition between opponents. This has traditionally been assumed to be unchanging—an eternal feature of war and human conflict. Colin Gray has written that "there is a unity to all strategic experience: nothing essential changes in the nature and function (or purpose) in sharp contrast to the character—of strategy and war."[1]

In his history of war in the nineteenth century, Azar Gat noted that "while the forms of war may change with time, its spirit or essence remains unchanged."[2] This theme recurs frequently in the writings of military and strategic theorists. Clausewitz defined the nature of war in the first instance through fighting. And at the core of his appreciation of war's nature was its reciprocity, which he expressed through metaphors such as the duel. War possesses its own dynamic or spirit, independent of its actors.[3]

On the other hand, the character of war is how wars are fought. War's character is accepted by military historians, as well as the members of our profession of arms, as constantly changing. The evolution of war is driven by factors such as culture, technology, politics, international relations, law, ethics, and military culture. Clausewitz wrote that "every era has its own kind of war, its own preconceptions."[4] This is no less true in our current era.

The character of war has irreversibly changed. This change has been occurring since humans started fighting and has occurred at a more rapid pace since the first Industrial Revolution. In the current era, the war for influence—largely a strategic fight—as well as the speed of operations, changes in international law and conventions, conceptions of ethical use of force, and new autonomous, precision, and long-range systems have driven this. Such change will continue.

However, some believe we are seeing the glimmerings of an evolution in the nature of war. War, at least until now, has been a human endeavor, driven (at least in the mind of Thucydides) by "fear, honor, interests." In Clausewitzian terms, the fog and friction of war have remained central to human conflict, as have human leadership and human agency.

In many instances, the proponents of a "changing nature" of war have confused its nature with its character. However, even more informed commentators have weighed in on this issue. In 2018 then–U.S. Secretary of Defense James Mattis speculated that a change in the nature of war was possible in the future: "I am questioning my original premise that the fundamental nature of war will not change." He told reporters in early 2018 that "if we ever get to the point where it is completely on automatic pilot, we are all spectators. That is no longer serving a political purpose. And conflict is a social problem that needs social solutions, people—human solutions. I'm certainly questioning my original premise that the fundamental nature of war will not change. You've got to question that now. I just don't have the answers yet."[5]

Frank Hoffman has posed the same question. In a 2018 article in *Parameters*, he examines how "modern-day heretics argue for capabilities in robotics, artificial intelligence, and human-machine teaming that will change more than just the way warfare is waged." He proposes that in the future, the nature of war may change due to the impact of these new technologies.[6]

Proponents of the "nature of war is changing" hypothesis base their argument primarily on the rapid developments in various forms of narrow artificial intelligence and autonomous systems in the last decade. These machines and algorithms have proliferated throughout our societies and into our military institutions. Some expect that machines and AI will be able to assume both the physical and cognitive functions of humans in war. If so, do we still call this undertaking "war"?

But we are still very much at the start of this race to deploy the highest quality and greatest quantity of robotic and semi-intelligent systems. Except for a small variety of systems that possess autonomy, such as shipboard close-in weapons systems, these robots and algorithms are still controlled by humans and subject to the operational concepts, organizational theories, and strategies designed and executed by human beings. While robotic systems are demonstrating an increasing capacity for a wider range of autonomous activities, they still lack the capacity to "understand" the broader context for their actions.

The current state of technological development has not seen the rise of intelligent machines. Algorithms used in artificial intelligence do not understand context. Without any appreciation of context or purpose—something humans do constantly—decision-making in war is unlikely to become the domain of intelligent machines. These machines, and AI, cannot and do not inspire or lead humans.

Nor have machines and algorithms attained the capacity for artificial general intelligence, which is broadly defined as the ability of a computer program to demonstrate human-like intelligent reasoning and decision-making—to do anything as well as, if not better than, humans. As a range of experts have described, we are still some distance from this goal. In Max Tegmark's book on AI, *Life 3.0*, experts concluded that the development of AGI may occur decades in the future at the earliest. Daniel Dennett has written that "the gap between today's systems and the science fiction systems dominating the popular imagination is still huge, though many folks, both lay and expert, manage to underestimate it."[7]

Toby Walsh has described four key obstacles in this quest for AGI: an inability to learn as quickly as humans, an inability to adapt, an inability to explain decisions (although many humans are also bad at this), and a lack of capacity to understand context or possess a deep understanding of our world.[8] Therefore, intelligent machines are some way off. Humans will retain the decision-making power over almost every aspect of warfare. If this is the case, the nature of war—which is ultimately a human contest of wills—endures.

We might simplify this argument further by posing a single question: Will humans still make the ultimate decision whether or not to go to war? It is difficult to imagine a future for humanity in which there is not some level of competition in the strategic environment and countries are not led by human politicians who make key strategic decisions (good and bad). War is therefore likely to remain a human endeavor, albeit one with more decision-support activities from artificial intelligence.

Hew Strachan has written that "if war remains an adversarial business whose dynamics create their own consequences, which can themselves be unpredictable, its nature cannot change."[9] Therefore, if you are in the camp that believes humans will continue to make the ultimate decisions about war, you probably believe that the nature of war remains unchanged.

If you don't believe this is the case, you may think that this situation is highly likely to change, and AI will eventually take over policy, strategy, and tactical actions related to war.

If you are not sure, you are probably in the majority!

My answer is that, for now, the nature of war remains unchanged. There is no technology with enough sophistication, creativity, or trustworthiness to make strategic decisions about high-risk human endeavors, especially war. There is no algorithm that can provide purpose for human activities and certainly no AI that can inspire humans to cohere, learn, fight, and sacrifice. While the pace of technological development is surging at present, no break-through in big data, high-performance computing, or artificial intelligence (and their associated disciplines) indicates that this situation will change in the coming decade or two.

# Notes

## Introduction

1. Sushant Singh, "Explained: If Soldiers on the LAC Were Carrying Arms, Why Did They Not Open Fire?" *The Indian Express*, 20 June 2020, https://indianexpress.com /article/explained/explained-if-soldiers-on-lac-were-carrying-arms-why-did-they-not -open-fire-6467324/.

2. Descriptions are drawn from contemporary press reports including Shubhajit Roy, "India-China Border Tension: Chinese Ambassador Acknowledges PLA Deaths," *The Indian Express*, 27 June 2020, https://indianexpress.com/article/india/india-china -border-tension-chinese-ambassador-acknowledges-pla-deaths-6478314/; Jeffrey Gettleman, Hari Kumar, and Sameer Yasir, "Worst Clash in Decades on Disputed India- China Border Kills 20 Indian Troops," *New York Times*, 16 June 2020; Yun Sun, "China's Strategic Assessment of the Ladakh Clash," *War on the Rocks*, 19 June 2020, https:// warontherocks.com/2020/06/chinas-strategic-assessment-of-the-ladakh-clash/.

3. Roberta Wohlstetter, *Pearl Harbor: Warning and Decision* (Stanford, CA: Stanford University Press, 1962), 397.

4. The earliest known examples are from Sun Tzu. In the West, Machiavelli published his "general rules" in *The Art of War*, and Clausewitz published his *Principles of War* in 1812.

5. Sun Tzu, *The Art of War*, trans. Ralph Sawyer (Boulder, CO: Westview Press, 1994), 168.

6. Ray Kurzweil and Chris Meyer, "Understanding the Accelerating Rate of Change," Kurzweilai.net blog, May 2003, https://www.kurzweilai.net/understanding-the -accelerating-rate-of-change; Will Steffen et al., "The Trajectory of the Anthropocene: The Great Acceleration," *The Anthropocene Review* 2, no. 1 (2015): 82–84.

7. Examples include Nora Bensahel and David Barno, *Adaptation under Fire: How Militaries Change in Wartime* (New York: Oxford University Press, 2020); Aimee Fox, *Learning to Fight: Military Innovation and Change in the British Army, 1914–1918* (Cambridge: Cambridge University Press, 2017); Williamson Murray and Allan Millett, eds., *Military Innovation in the Interwar Period* (Cambridge: Cambridge University Press, 1996); Dima Adamsky, *The Culture of Military Innovation: The Impact of Cultural Factors on the Revolution in Military Affairs in Russia, the U.S., and Israel* (Stanford, CA: Stanford Security Studies, 2010); Stephen Rosen, *Winning the Next War: Innovation and the Modern Military* (Ithaca: Cornell University Press, 1991); and Stephen Biddle, *Military Power: Explaining Victory and Defeat in Modern Battle* (Princeton: Princeton University Press, 2006).

8. Paul Kennedy, *The Rise and Fall of the Great Powers: Economic Change and Military Conflict from 1500 to 2000* (New York: Random House, 1997), 191–92.

9. For a detailed account of Napoleon's defeat and occupation of Prussia in 1806, see David Chandler, *The Campaigns of Napoleon: The Mind and Method of History's Greatest Soldier* (New York: Scribner, 1966), 443–551, and Peter Paret, *The Cognitive Challenge of War: Prussia 1806* (Princeton: Princeton University Press, 2009). The battle of Tsushima in May 1905 saw a Japanese fleet achieve a decisive victory over the Russians. It was the first major battle fought by modern steel battleships and the first in which wireless telegraphy played a major role. John Keegan, *A History of Warfare* (New York: Alfred A. Knopf Inc., 1993), 67–68; and Niall Ferguson, *The War of the World: Twentieth-Century Conflict and the Descent of the West* (New York: Penguin Books, 2006), 55–56. The French defeat in 1940 is explored in Alistair Horne, *To Lose a Battle* (London: Penguin Books, 1979), and Julian Jackson, *The Fall of France: The Nazi Invasion of 1940* (Oxford: Oxford University Press, 2004).

10. Competitive strategies are those adopted by a nation for the peacetime use of latent military power to shape the choices of a competitor in ways that favor the nation implementing the competitive strategy. Thomas Mahnken, ed., *Competitive Strategies for the 21st Century: Theory, History, and Practice* (Stanford, CA: Stanford University Press, 2012), 5–7; also, Andrew Marshall, "Competitive Strategies: History and Background," unpublished paper (Washington, DC: Department of Defense, 3 March 1988).

11. Audrey Kurth Cronin, *Power to the People: How Open Technological Innovation Is Arming Tomorrow's Terrorists* (New York: Oxford University Press, 2020), 4.

12. John Lynn, *Battle: A History of Combat and Culture* (New York: Westview Press, 2003).

13. Thomas Mahnken, *Technology and the American Way of War since 1945* (New York: Columbia University Press, 2008), 225.

14. MacGregor Knox and Williamson Murray, eds., *The Dynamics of Military Revolution, 1300–2050* (Cambridge: Cambridge University Press, 2001), 179–80.

15. Biddle, 190–91.

16. Andrew Marshall, "Some Thoughts on Military Revolutions (Second Version), Memorandum for the Record" (Washington, DC: Department of Defense, 23 August 1993), 2.

17. Hew Strachan, *Clausewitz's* On War*: A Biography* (New York: Grove Press, 2007), 193.

18. Azar Gat, *War in Human Civilization* (Oxford: Oxford University Press, 2006), 663.

19. Book 1 of Thucydides' *The Peloponnesian War*, in *The Landmark Thucydides: A Comprehensive Guide to the Peloponnesian War*, ed. Robert Strassler (New York: Free Press, 1998). Also, Donald Kagan, "Intangible Interests and U.S. Foreign Policy," *Commentary* (April 1997), https://www.cs.utexas.edu/users/vl/notes/kagan.html.

20. Donald Kagan, *The Peloponnesian War* (London: Penguin Books, 2004).

21. Alan Moorehead, *Gallipoli* (Melbourne: Macmillan Company of Australia, 1989), and Charles Bean, *Official History of Australia in the War of 1914–1918*, vol. 1: *The Story of ANZAC* (Sydney: Angus and Robertson Ltd., 1941).

22. As Clausewitz noted, war is an act of force to compel our enemy to do our will. Carl von Clausewitz, *On War*, trans. Michael Howard and Peter Paret (Princeton: Princeton University Press, 1989), 75.

23. The original work is lost; however, elements of this work remain in his most famous work, *Histories*.

24. Edward MacCurdy, ed., *The Notebooks of Leonardo da Vinci* (Old Saybrook, CT: Konecky and Konecky, 1955), 806.

25. Victor Davis Hanson, *The Western Way of War: Infantry Battle in Classical Greece* (Berkeley: University of California Press, 2009).

26. J. F. C. Fuller, *The Generalship of Alexander the Great* (Hertfordshire, UK: Wordsworth Editions Limited, 1998), 47.

27. Philip Freeman, *Alexander the Great* (New York: Simon and Schuster, 2011).

28. Chandler, 138–42.

29. A little over a year after the 1793 proclamation of the levée en masse, France had 1,169,000 soldiers under arms. Keegan, 233.

30. This convergence of new organizations, new technology, and new warfighting concepts features in the work of Williamson Murray and Allan Millett, Andrew Krepinevich, and Stephen Biddle, among others.

31. Knox and Murray, 12.

32. Samuel Huntington, *The Soldier and the State: The Theory and Politics of Civil-Military Relations* (Cambridge, MA: Belknap Press, 1957); Morris Janowitz, *The Professional Soldier: A Social and Political Portrait* (New York: Free Press, 1964).

33. In their book on military failure, Cohen and Gooch examine key drivers of failure: failure to learn, failure to anticipate, failure to adapt, and aggregate failure. Eliot Cohen and John Gooch, *Military Misfortunes: The Anatomy of Failure in War* (New York: Vintage Books, 1991).

34. Williamson Murray, *War, Strategy, and Military Effectiveness* (Cambridge: Cambridge University Press, 2011), 3.

## Chapter 1. Revolutions and Military Change

1. Steven Levy, "What Deep Blue Tells Us about AI in 2017," *Wired Magazine*, 23 May 2017, https://www.wired.com/2017/05/what-deep-blue-tells-us-about-ai-in-2017/.

2. Larry Greenemeier, "20 Years after Deep Blue: How AI Has Advanced since Conquering Chess," *Scientific American,* 2 June 2017, https://www.scientificamerican.com/article/20-years-after-deep-blue-how-ai-has-advanced-since-conquering-chess/.

3. "IBM's 100 Icons of Progress: Deep Blue," IBM.com, https://www.ibm.com/ibm/history/ibm100/us/en/icons/deepblue/.

4. Levy.

5. "Kasparov vs. Deep Blue: The Match That Changed History," Chess.com, 12 October 2018, https://www.chess.com/article/view/deep-blue-kasparov-chess.

6. Knox and Murray, 12–13.

7. Thomas Crump, *The Age of Steam: The Power That Drove the Industrial Revolution* (London: Robinson, 2007), 50–58.

8. William Bernstein, *The Birth of Plenty: How the Prosperity of the Modern World Was Created* (New York: McGraw-Hill, 2004), 18–20.

9. Williamson Murray and Wayne Hsieh, *A Savage War: A Military History of the Civil War* (Princeton: Princeton University Press, 2016).

10. Vaclav Smil, *Creating the Twentieth Century: Technical Innovations of 1867–1914 and Their Lasting Impact* (Oxford: Oxford University Press, 2005), 4.

11. Smil, 9.

12. Andrew Liaropoulos, "Revolutions in Warfare: Theoretical Paradigms and Historical Evidence: The Napoleonic and First World War Revolutions in Military Affairs," *The Journal of Military History* 70, no. 2 (April 2006): 363–84; Knox and Murray; and Murray and Millett.

13. Max Boot, *War Made New: Technology, Warfare, and the Course of History, 1500 to Today* (New York: Gotham Books, 2006), 14.

14. Smil, 303.

15. The origins and development of the Internet have been described in multiple sources, including James Gillies and Robert Cailliau, *How the Web Was Born: The Story of the World Wide Web* (Oxford: Oxford University Press, 2000); Katie Hafner and Matthew Lyon, *Where Wizards Stay Up Late: The Origins of the Internet* (New

York: Simon and Schuster, 1998); and Tim Berners-Lee, *Weaving the Web: The Original Design and Ultimate Destiny of the World Wide Web* (New York: Harper Business, 2000). Also Smil, 4.

16. Klaus Schwab, *The Fourth Industrial Revolution* (New York: Crown Business, 2016).

17. Knox and Murray, 4; Marshall, "Some Thoughts on Military Revolutions," 4–5.

18. Gerhard Mensch, *Stalemate in Technology: Innovations Overcome the Depression* (Cambridge: Ballinger, 1979), 135.

19. Kurzweil and Meyer.

20. Thomas Friedman, *Thank You for Being Late* (New York: Macmillan, 2016). Moore's Law is a computing term that states that processor speeds or overall processing power for computers will double every two years.

21. Friedman, 63–66.

22. Steffen et al., "The Trajectory of the Anthropocene," 82.

23. Will Steffen et al., *Global Change and the Earth System: A Planet under Pressure* (Berlin: IGBP Book Series, 2004), 131.

24. Intergovernmental Panel on Climate Change (IPCC), *Climate Change 2014: Synthesis Report Summary for Policy Makers* (Geneva: IPCC, 2015), 2.

25. Boot, 16.

26. National Intelligence Council (NIC), *Global Trends 2010* (Washington, DC: NIC, 1997); NIC, *Mapping the Global Future* (Washington, DC: NIC, 2004), 34; U.S. Department of Defense (DOD), *Sustaining U.S. Global Leadership: Priorities for 21st Century Defense* (Washington, DC: DOD, 2012), 1.

27. NIC, *Global Trends 2017: Paradox of Progress* (Washington, DC: NIC, 2017), 177.

28. U.S. Army Training and Doctrine Command, *The Operational Environment and the Changing Character of Warfare* (Fort Eustis, VA: Department of the Army, October 2019).

29. Philipp Blom, *The Vertigo Years: Change and Culture in the West, 1900–1914* (London: Phoenix Books, 2009), 2–3.

30. Edward Bryn, *The Progress of Invention in the Nineteenth Century* (New York: Munn and Co. Publishers, 1900), 466.

31. Mensch, 160–62.

32. Tim Unwin, "5 Problems with the 4th Industrial Revolution," *ICTWorks*, 23 March 2019, https://www.ictworks.org/problems-fourth-industrial-revolution/#.X97vMC0Rokh.

33. Elizabeth Garbee, "This Is Not the Fourth Industrial Revolution," Slate.com, 29 January 2016, https://slate.com/technology/2016/01/the-world-economic-forum-is-wrong-this-isnt-the-fourth-industrial-revolution.html.

34. Smil, 4–6.

35. George Orwell, BBC radio broadcast on 10 March 1942, in Smil.

36. Blom, 2–3.

37. Chris McKenna, "Our World Is Changing—But Not as Rapidly as People Think," World Economic Forum, 2 August 2018, https://www.weforum.org/agenda/2018/08/change-is-not-accelerating-and-why-boring-companies-will-win/.

38. Knox and Murray, 11–14.

39. Michael O'Hanlon, *Forecasting Change in Military Technology, 2020–2040* (Washington, DC: Brookings Institution, September 2018), 4.

40. The Atlantic Council owns a Foresight, Strategy, and Risk Initiative; see https://www.atlanticcouncil.org/programs/scowcroft-center-for-strategy-and-security/foresight-strategy-and-risks-initiative/. RAND maintains a Center for Futures and Foresight Studies at https://www.rand.org/randeurope/methods/futures-and-foresight-studies.html.

41. Francis Fukuyama, "The End of History?" *The National Interest* no. 16 (Summer 1989), 3–4.

42. Francis Fukuyama, *The End of History and the Last Man* (New York: Avon Books, 1992).

43. Fukuyama, *The End of History and the Last Man*, xii–xiii.

44. Timothy Stanley and Alexander Lee, "It's Still Not the End of History," *The Atlantic*, 1 September 2014, https://www.theatlantic.com/politics/archive/2014/09/its-still-not-the-end-of-history-francis-fukuyama/379394/.

45. Starting in 2006, *The Economist Intelligence Unit* has kept a running watch on developments in democratic governance around the world. Its 2019 Democracy Index gave the lowest rating for democracy since the survey commenced. A key finding was that only 22 nations, with a combined population of 430 million people, could be described as "full democracies." The U.S.-based Freedom House found in its *Freedom in the World 2019* report that there have been thirteen consecutive years of decline in political rights and civil liberties around the world. "Global Democracy Has Another Bad Year," *The Economist*, 22 January 2020; Freedom House, *Freedom in the World 2019* (Washington, DC: Freedom House, 2019), 1.

46. Larry Diamond, "Facing Up to the Democratic Recession," *Journal of Democracy* 26, no. 1 (January 2015): 141–55.

47. Robert Kagan, "The Twilight of the Liberal World Order," Brookings Institution blog, 24 January 2017, https://www.brookings.edu/research/the-twilight-of-the-liberal-world-order/.

48. Michael Mazarr et al., *Understanding the Emerging Era of International Competition: Theoretical and Historical Perspectives* (Santa Monica, CA: RAND Corporation, 2018), 1.

49. NIC, *Global Trends 2017*, 35.

50. U.S. Department of Defense, *Military and Security Developments Involving the People's Republic of China 2018* (Washington, DC: Office of the Secretary of Defense, 2019), 43–46.

51. Robert Seely, "Defining Contemporary Russian Warfare," *RUSI Journal* 162, no. 1 (2017): 50–59.

52. Thomas Mahnken, Ross Babbage, and Toshi Yoshihara, *Countering Comprehensive Coercion: Strategies Against Authoritarian Political Warfare* (Washington, DC: Center for Defense and Strategic Studies, 30 May 2018), 1–8.

53. Speech by General Valery Gerasimov to the Russian Academy of Military Sciences, 2 March 2019. A translation of key elements of the speech is contained in Roger McDermott, "Gerasimov Unveils Russia's Strategy of Limited Actions," *Eurasia Daily Monitor*, 6 March 2019, https://jamestown.org/program/gerasimov-unveils-russias -strategy-of-limited-actions/.

54. Bruce Jones, "The New Geopolitics," Brookings Institution blog, 28 November 2017, https://www.brookings.edu/blog/order-from-chaos/2017/11/28/the-new-geopolitics/.

55. Mahnken et al., 30–42.

56. Oriana Skylar Mastro, "The Stealth Superpower: How China Hid Its Global Ambitions," *Foreign Affairs* 98, no. 1 (January/February 2019), 32.

57. Ross Babbage, *Winning without Fighting: Chinese and Russian Political Warfare Campaigns and How the West Can Prevail*, vol. 1 (Washington, DC: Center for Strategic and Budgetary Assessments, 2019), i.

58. Maria Snegovaya, *Putin's Information Warfare in Ukraine: Soviet Origins of Hybrid Warfare*, Russia Report 1 (Washington, DC: Institute for the Study of War, September 2015), 7.

59. Babbage, *Winning without Fighting*, 46.

60. Andrew Chatzky and James McBride, "China's Massive Belt and Road Initiative," Council on Foreign Relations Backgrounder, 28 January 2020, https://www.cfr .org/backgrounder/chinas-massive-belt-and-road-initiative.

61. Mastro, 32.

62. Ellen Ioanes, "China Steals U.S. Designs for New Weapons, and It's Getting Away with the Greatest Property Theft in Human History," *Business Insider*, 25 September 2019, https://www.businessinsider.com.au/esper-warning-china-intellectual-property-theft -greatest-in-history-2019-9?r=US&IR=T.

63. Wayne Morrison, *The Made in China 2025 Initiative: Economic Implications for the United States* (Washington, DC: Congressional Research Service, 12 April 2019).

64. *Made in China 2025: Backgrounder 2018* (Stockholm: Institute for Security and Development Policy, June 2018), https://isdp.eu/content/uploads/2018/06/Made-in -China-Backgrounder.pdf.

65. William Overholt, *China's Crisis of Success* (Cambridge: Cambridge University Press, 2018).

66. Susan Shirk, *Fragile Superpower: How China's Internal Politics Could Derail Its Peaceful Rise* (Oxford: Oxford University Press, 2007).

67. Michael Chase et al., *China's Incomplete Military Transformation: Assessing the Weaknesses of the People's Liberation Army (PLA)* (Santa Monica, CA: RAND Corporation, 2015), 135–39.

68. According to the U.S. Census Bureau in 2019, India's population was 1.3 billion people, and Pakistan's was 210 million people. U.S. Census Bureau, "U.S. and World Population Clock," 8 February 2019, https://www.census.gov/popclock/.

69. DOD, *The Joint Force in a Contested and Disordered World* (Washington, DC: Joint Staff, 14 July 2016), ii; DOD, *National Defense Strategy of the United States of America* (Washington, DC: DOD, 2017), 2–3.

70. The reasons for the inadequacy of these ideas for the digital age strategic competition are explained in Eliot Cohen, *The Big Stick: The Limits of Soft Power and the Necessity of Military Force* (New York: Basic Books, 2016), 200–206.

71. Alan Dupont, *Mitigating the New Cold War: Managing U.S.-China Trade, Tech, and Geopolitical Conflict* (Sydney: Centre for Independent Studies, 2020), 1.

72. Thomas Malthus, *An Essay on the Principle of Population as It Affects the Future Improvement of Society* (London: J. Johnson Publishers, 1798).

73. Malthus, 26.

74. United Nations (UN), *World Population Prospects 2019: Highlights* (New York: UN Department of Economic and Social Affairs, Population Division, 2019), 1.

75. Wolfgang Lutz, "World Population Trends and the Rise of Homo Sapiens Literata," Working Paper 19-012 (Laxenburg, Austria: International Institute for Applied Systems Analysis, December 2019), 13–15, http://pure.iiasa.ac.at/id/eprint/16214/1/WP-19-012.pdf.

76. Darryl Bricker and John Ibbitson, *Empty Planet: The Shock of Global Population Decline* (New York: Crown Publishers, 2019), 2.

77. Bricker and Ibbitson, 50–54.

78. UN, *World Population Aging 2019* (New York: UN Department of Economic and Social Affairs, Population Division, 2020), 1.

79. UN, *World Population Aging 2019*, 1–5.

80. UN, *World Urbanization Prospects: The 2018 Revision* (New York: UN Department of Economic and Social Affairs, 16 May 2018); UN, *International Migration Report 2017* (New York: UN Department of Economic and Social Affairs, Population Division, 2017), 1; Roman Muzalevsky, *Strategic Landscape 2050* (Carlisle, PA: Strategic Studies Institute and U.S. Army War College Press, 2017), 12.

81. Jonathan Woetzel et al., *People on the Move: Global Migration's Impact and Opportunity* (New York: McKinsey Global Institute, December 2016).

82. UN, *World Migration Report 2020* (New York: UN Department of Economic and Social Affairs, Population Division, 2019), 19.

83. The role of technology in the imperial expansion of European nations between 1415 and 1914 is described in a range of sources. One particularly good account is Ian Morris, *War: What Is It Good For? The Role of Conflict in Civilization from Primates to Robots* (London: Profile Books, 2014).

84. Knox and Murray, 36–53.

85. Milton Hoenig, "Artificial Intelligence: A Detailed Explainer, with a Human Point of View," *Bulletin of the Atomic Scientists*, 7 December 2018, https://thebulletin.org/2018/12/artificial-intelligence-a-detailed-explainer-with-a-human-point-of-view/.

86. Future of Life Institute, "The Benefits and Risks of Artificial Intelligence," Cambridge University, https://futureoflife.org/background/benefits-risks-of-artificial-intelligence/?cn-reloaded=1.

87. Kenneth Payne, "Artificial Intelligence: A Revolution in Strategic Affairs?" *Survival* 60, no. 5 (October–November 2018): 23.

88. Colin Gray, *The Future of Strategy* (Cambridge: Polity Press, 2015), 117.

89. James Holmes and Toshi Yoshihara, *Red Star Over the Pacific: China's Rise and the Challenge to U.S. Maritime Strategy*, 2nd ed. (Annapolis, MD: Naval Institute Press, 2018), 23.

90. Paul Scharre, "Are AI-Powered Killer Robots Inevitable?" *Wired*, 19 May 2020, https://www.wired.com/story/artificial-intelligence-military-robots/.

91. Heather Roff, *Uncomfortable Ground Truths: Predictive Analytics and National Security* (Washington, DC: Brookings Institution, 2020), 5–6.

92. Max Boot examines this idea of employing robots for the "dirty, dangerous, and dull" tasks in *War Made New*, 442.

93. The Strategic Capabilities Office, partnering with Naval Air Systems Command, demonstrated the micro-drone swarms at China Lake, California. The test, conducted in October 2016 and documented on the CBS News program *60 Minutes*, comprised 103 Perdix drones launched from three F/A-18 Super Hornets. DOD media release, "Department of Defense Announces Successful Micro-Drone Demonstration," 9 January 2017, https://www.defense.gov/News/News-Releases/News-Release-View/Article/1044811/department-of-defense-announces-successful-micro-drone-demonstration/.

94. Joseph Trevithick, "Huge Navy Unmanned-Focused Experiment Underway Featuring Live Missile Shoot and Super Swarms," *The Warzone*, 20 April 2021, https://www.thedrive.com/the-war-zone/40262/huge-navy-unmanned-focused-experiment-underway-featuring-live-missile-shoot-and-super-swarms.

95. Joseph Trevithick, "China Conducts Test of Massive Suicide Drone Swarm Launched from a Box on a Truck," *The Warzone*, 14 October 2020, https://www.thedrive.com/the-war-zone/37062/china-conducts-test-of-massive-suicide-drone-swarm-launched-from-a-box-on-a-truck.

96. Michael Ryan, *Human Machine Teaming for Future Ground Forces* (Washington, DC: Center for Strategic and Budgetary Assessments, 25 April 2018), 14–15.

97. Sydney Freedberg, "Meet the Army's Future Family of Robot Tanks: RCV," *Breaking Defense*, 9 November 2020, https://breakingdefense.com/2020/11/meet -the-armys-future-family-of-robot-tanks-rcv/.

98. Peter Suciu, "China's Army Now Has Killer Robots: Meet the Sharp Claw," *National Interest* blog, 17 April 2020, https://nationalinterest.org/blog/buzz/chinas -army-now-has-killer-robots-meet-sharp-claw-145302.

99. Ryan, *Human Machine Teaming for Future Ground Forces*, 14–15.

100. Zachery Kallenborn, "Autonomous Drone Swarms as WMD," Modern War Institute, 28 May 2020, https://mwi.usma.edu/swarms-mass-destruction-case-declaring -armed-fully-autonomous-drone-swarms-wmd/.

101. Scharre, "Are AI-Powered Killer Robots Inevitable?"

102. Jonathan Dowling and Gerard Milburn, "Quantum Technology: The Second Quantum Revolution," *The Royal Society* 361 (2003): 1656–57; Elsa Kania and John Costello, *Quantum Hegemony: China's Ambitions and the Challenge to U.S. Innovation Leadership* (Washington, DC: Center for a New American Security, 2018), 2–5.

103. Tom Stefanick, "The State of U.S.-China Quantum Data Security Competition" (Washington, DC: Brookings Institution, 18 September 2020).

104. Elsa Kania, "China's Quantum Future: Xi's Quest to Build a High-Tech Superpower," *Foreign Affairs*, 26 September 2018, https://www.foreignaffairs.com /articles/china/2018-09-26/chinas-quantum-future.

105. Lindsay Rand and Berit Goodge, "Information Overload: The Promise and Risk of Quantum Computing," *Bulletin of the Atomic Scientists*, 14 November 2019, https://thebulletin.org/2019/11/information-overload-the-promise-and-risk-of -quantum-computing/; Michael Biercuk and Richard Fontaine, "The Leap into Quantum Technology: A Primer for National Security Professionals" *War on the Rocks*, 17 November 2017, https://warontherocks.com/2017/11/leap-quantum -technology-primer-national-security-professionals/.

106. Kania and Costello, 14.

107. Kania and Costello, 5.

108. Michael Vermeer and Evan Peet, *Securing Communications in the Quantum Computing Age* (Santa Monica, CA: RAND Corporation, 2020), 1.

109. Jeremy Berg, John Tymoczko, and Lubert Stryer, *Biochemistry* (New York: W. H. Freeman, 2002).

110. National Institutes of Health, "DNA Sequencing Costs: Data," https://www.genome .gov/27541954/dna-sequencing-costs-data/.

111. Jennifer Doudna et al., "A Programmable Dual-RNA-Guided DNA Endonuclease in Adaptive Bacterial Immunity," *Science*, 17 August 2012, 816–21; Laura Kahn,

"A Crispr Future," *Bulletin of the Atomic Scientists*, 16 December 2015, https://thebulletin.org/2015/12/a-crispr-future/.

112. Charlotte Jee, "The First U.S. Trial of CRISPR Gene Editing in Cancer Patients Suggests the Technique Is Safe," *MIT Tech Review*, 7 February 2020, https://www.technologyreview.com/f/615157/the-first-us-trial-of-crispr-gene-editing-in-cancer-patients-suggests-the-technique-is-safe/.

113. Anne Trafton, "Making Smart Materials with CRISPR," *MIT Tech Review*, 24 October 2019, https://www.technologyreview.com/s/614509/making-smart-materials-with-crispr/.

114. Defense Advanced Research Projects Agency, "High Energy Liquid Laser Area Defense System," https://www.darpa.mil/program/high-energy-liquid-laser-area-defense-system.

115. Dan Robitzski, "U.S. Army Doubles Down on Directed Energy Weapons," *Futurism.com*, 2 August 2019, https://futurism.com/the-byte/army-directed-energy-weapons.

116. Lockheed Martin, "Directed Energy," https://www.lockheedmartin.com/en-us/capabilities/directed-energy.html.

117. Ben Warner, "Pentagon Shifts Focus on Directed Energy Weapons Technology," U.S. Naval Institute blog, 5 September 2019, https://news.usni.org/2019/09/05/pentagon-shifts-focus-on-directed-energy-weapons-technology.

118. "Russia's New MiG-35 Fighter Jet to Use Laser Weapons," *Pravda.Ru*, 27 January 2017, https://www.pravdareport.com/news/russia/136730-mig_35_laser/.

119. Jeffrey Lin and Peter Singer, "Drones, Lasers, and Tanks: China Shows Off Its Latest Weapons," *Popular Science*, 27 February 2017, https://www.popsci.com/china-new-weapons-lasers-drones-tanks/.

120. Eric Heginbotham et al., *The U.S.-China Military Scorecard* (Santa Monica, CA: RAND Corporation, 2017), 246–48.

121. U.S. Department of Defense, *Military and Security Developments Involving the People's Republic of China 2019* (Washington, DC: Office of the Secretary of Defense, 2019), 56.

122. MBDA Missile Systems, "Dragonfire Laser Turret Unveiled at DSEI 2017," 12 September 2017, https://www.mbda-systems.com/press-releases/dragonfire-laser-turret-unveiled-dsei-2017/.

123. Richard Speier et al., *Hypersonic Missile Nonproliferation: Hindering the Spread of a New Class of Weapons* (Santa Monica, CA: RAND Corporation, 2017).

124. International Institute for Strategic Studies, "Hypersonic Weapons and Strategic Stability," *Strategic Comments* 26 (March 2020), 1.

125. David Vergun, "Military Leaders Discuss Hypersonics, Supply Chain Vulnerabilities," *DoD News*, 21 February 2020.

126. Speier et al., xii.

127. Speier et al., 13.

128. Carl Sagan, quoted in Robert Zubrin, *Entering Space: Creating a Spacefaring Civilization* (New York: Penguin Putnam Inc., 1999), ix.

129. See https://www.starlink.com.

130. DOD, *Military and Security Developments Involving the People's Republic of China 2019*, 50–51; Phil Stewart, "U.S. Studying India Anti-Satellite Weapons Test, Warns of Space Debris," *Reuters*, 28 March 2019.

131. Anatoly Zak, "The Hidden History of the Soviet Satellite Killer," *Popular Mechanics*, 1 November 2013, https://www.popularmechanics.com/space/satellites/a9620 /the-hidden-history-of-the-soviet-satellite-killer-16108970/.

132. John Sargent and R. X. Schwartz, *3D Printing: Overview, Impacts, and the Federal Role* (Washington, DC: Congressional Research Service, 2 August 2019), 6–7.

133. Sargent and Schwartz, 8.

134. T. X. Hammes, "3D Printing Will Disrupt the World in Ways We Can Barely Imagine," *War on the Rocks*, 28 December 2015, https://warontherocks.com /2015/12/3-d-printing-will-disrupt-the-world-in-ways-we-can-barely-imagine/.

135. See UPS, "3D Printing Services," https://www.ups.com/us/en/services/high-tech /3d-print.page.

136. "Special Report: The Facts about Over Consumption," *New Scientist*, 15 October 2008, https://www.newscientist.com/article/dn14950-special-report-the-facts-about -overconsumption/; Steffen et al., "The Trajectory of the Anthropocene," 84–88.

137. Steffen et al., "The Trajectory of the Anthropocene," 82–84.

138. Paul Crutzen and Eugene Stoermer, "The Anthropocene," *IGBP Newsletter*, no. 41, May 2000, 17, http://www.igbp.net/download/18.316f18321323470177580001401 /1376383088452/NL41.pdf.

139. Anthropocene Working Group, Sub-commission on Quaternary Stratigraphy, International Union of Geological Sciences, "Working Group on the Anthropocene," 21 May 2019, http://quaternary.stratigraphy.org/working-groups/anthropocene/.

140. IPCC, *Climate Change 2014: Synthesis Report. Contribution of Working Groups I, II, and III to the Fifth Assessment Report of the Intergovernmental Panel on Climate Change* (Geneva: IPCC, 2014), 4.

141. IPCC, *Climate Change 2014*, 5.

142. Kyle Harper, "The Coronavirus Is Accelerating History Past the Breaking Point," *Foreign Policy*, 6 April 2020, https://foreignpolicy.com/2020/04/06/coronavirus -is-accelerating-history-past-the-breaking-point/.

143. This is a key theme in Harper.

144. Mahnken, *Competitive Strategies for the 21st Century*, 12.

145. U.S. Marine Corps, *Competing* (Washington, DC: Headquarters Marine Corps, 14 December 2020), 1–3.

146. The captain of a Soviet submarine had to be talked out of using a nuclear missile by his crew. Michael Dobbs, *One Minute to Midnight* (New York: Knopf, 2008), 317.

147. Muzalevsky, 12.

148. The U.S. National Security Strategy released in 2017 describes an environment of reemerging strategic competition and the range of challenges to its capacity to deter aggression. DOD, *National Security Strategy of the United States of America* (Washington, DC: DOD, December 2017), 2–3; U.S. Government, *Interim National Security Strategic Guidance* (Washington, DC: White House, 2021), 7–8; Government of Japan, *Defense of Japan 2017* (Tokyo: Government of Japan, 2017), 207–9, http://www.mod.go.jp/e/publ/w_paper/2017.html; Government of Japan, *Defense of Japan 2018* (Tokyo: Government of Japan, 2018), http://www.mod.go.jp/e/publ/pdf/2018/DOJ2018_Digest_0827.pdf.

149. Colin Gray, "War—Continuity in Change, and Change in Continuity," *Parameters* 40, no. 2 (Summer 2010): 12.

150. United Kingdom Ministry of Defence, *Global Strategic Trends: The Future Starts Today*, 6th ed. (Swindon: Development, Concepts and Doctrine Centre, 2018), 31–34.

151. Daniel R. Coats, "Worldwide Threat Assessment of the U.S. Intelligence Community," Statement for the Record to Senate Select Committee on Intelligence (Washington, DC: Office of the Director of National Intelligence, 29 January 2019), 23; Office of the Director of National Intelligence, *Annual Threat Assessment of the U.S. Intelligence Community* (Washington, DC: Office of the Director of National Intelligence, 9 April 2021), 18.

## Chapter 2. Future War

1. Qiao Liang and Wang Xiangsui, *Unrestricted Warfare* (Beijing: PLA Literature and Arts Publishing House, February 1999), 7.

2. Tota Ishimaru, *The Next World War* (London: Hurst and Blackett, 1936), 36.

3. Ishimaru, 340.

4. Ishimaru, 343.

5. Lawrence Freedman, *The Future of War: A History* (New York: Public Affairs Press, 2017), ix.

6. Norman Angell, *The Great Illusion: A Study of the Relation of Military Power in Nations to their Economic and Social Advantage* (London: Putnam and Sons, 1911).

7. Colin Gray, *Strategy and Defence Planning: Meeting the Challenge of Uncertainty* (Oxford: Oxford University Press, 2014), 94.

8. Frank Hoffman, "Will War's Nature Change in the Seventh Military Revolution?" *Parameters* 47, no. 7 (Winter 2017–18): 23.

9. Colin Gray, *Modern Strategy* (Oxford: Oxford University Press, 1999), 362.

10. Clausewitz, 578.

11. Strassler, *The Landmark Thucydides*, Book 1; Kagan, "Intangible Interests and U.S. Foreign Policy."

12. Michael Howard, *The Causes of War*, 2nd ed. (Cambridge, MA: Harvard University Press, 1983), 25.

13. DOD, *Indo-Pacific Strategy Report* (Washington, DC: DOD, June 2019), 2–7.

14. Biddle, ix.

15. Gat, *War in Human Civilization*, 663.

16. Morris, *War: What Is It Good For?*, 391.

17. Bethany Lacina and Nils Gleditsch, "The Waning of War Is Real: A Response to Gohdes and Price," *Journal of Conflict Resolution* 57, no. 6 (2012): 1109–27; Bethany Lacina and Nils Gleditsch, "Monitoring Trends in Global Combat: A New Dataset of Battle Deaths," *European Journal of Population* 21 (2005): 145–66; John Mueller, "War Has Almost Ceased to Exist," *Political Science Quarterly* 124, no. 2 (Summer 2009): 297–321.

18. Steven Pinker, *The Better Angels of Our Nature: Why Violence Has Declined* (New York: Penguin Books, 2011).

19. Pinker, 31–58.

20. Pinker, 59–85.

21. Pinker, 128–88.

22. Pinker credits the title of this chapter to John Lewis Gaddis's *The Long Peace: Inquiries into the History of the Cold War*. Pinker, 189–294.

23. Pinker, 295–377.

24. Pinker, 378–481.

25. R. Brian Ferguson, "Pinker's List: Exaggerating Prehistoric War Mortality," in *War, Peace, and Human Nature: The Convergence of Evolutionary and Cultural Views*, ed. Douglas Fry (Oxford: Oxford University Press, 2013), 126.

26. Examples include Elizabeth Kolbert, "Peace in Our Time: Steven Pinker's History of Violence," *The New Yorker*, 26 September 2011, https://www.newyorker.com /magazine/2011/10/03/peace-in-our-time-elizabeth-kolbert;andJohnArquilla,"The Big Kill," *Foreign Policy*, 3 December 2012, https://foreignpolicy.com/2012/12/03/the -big-kill/.

27. Pasquale Cirillo and Nassim Taleb, "On the Statistical Properties and Tail Risk of Violent Conflicts," *Physica A: Statistical Mechanics and its Applications* 452 (2016): 29–45.

28. Bear Braumoeller, *Only the Dead: The Persistence of War in the Modern Age* (New York: Oxford University Press, 2019).

29. Braumoeller, 8.

30. Emmanuel Gfoeller, "Gap Warfare: The Case for a Shift in America's Strategic Mindset," *Real Clear Defense*, 28 May 2020, https://www.realcleardefense.com/articles /2020/05/28/gap_warfare_the_case_for_a_shift_in_americas_strategic_mindset _115325.html.

31. Donald Stoker and Craig Whiteside, "Blurred Lines: Gray-Zone Conflict and Hybrid War—Two Failures of American Strategic Thinking," *Naval War College Review* 73, no. 1 (2020): 13.

32. Clausewitz, 146.

33. Eliot Cohen, *Supreme Command: Soldiers, Statesmen, and Leadership in Wartime* (New York: Free Press, 2002), 208.

34. Beatrice Heuser, *The Evolution of Strategy: Thinking War from Antiquity to the Present* (Cambridge: Cambridge University Press, 2010), 485.

35. Trevor Dupuy, *The Evolution of Weapons and Warfare* (New York: Da Capo Press, 1984), 287.

36. The issue of resistance to change in military organizations is examined in a range of books and other publications. Among the best are: Stephen Rosen, *Winning the Next War: Innovation and the Modern Military* (Ithaca: Cornell University Press, 1991); James Corum, *The Roots of Blitzkrieg* (Lawrence: University Press of Kansas, 1992); Williamson Murray and Allan Millett, eds., *Military Innovation in the Interwar Period* (Cambridge: Cambridge University Press, 1996); MacGregor Knox and Williamson Murray, *The Dynamics of Military Revolution, 1300–2050* (Cambridge: Cambridge University Press, 2001); David Johnson, *Fast Tanks and Heavy Bombers: Innovation in the U.S. Army, 1917–1945* (Ithaca: Cornell University Press, 2003); Michael O'Hanlon, *Technological Change and the Future of Warfare* (Washington, DC: Brookings Institution, 2000); Dima Adamsky, *The Culture of Military Innovation: The Impact of Cultural Factors on the Revolution in Military Affairs in Russia, the U.S., and Israel* (Stanford, CA: Stanford University Press, 2010); and Frank Hoffman, *Mars Adapting: Military Change during War* (Annapolis, MD: Naval Institute Press, 2021).

37. Elizabeth Kier, *Imagining War: French and British Military Doctrine between the Wars* (Princeton: Princeton University Press, 1997), 32.

38. David Johnson, "An Army Caught in the Middle between Luddites, Luminaries, and the Occasional Looney," *War on the Rocks*, 19 December 2018, https://warontherocks .com/2018/12/an-army-caught-in-the-middle-between-luddites-luminaries-and-the -occasional-looney/.

39. Their three-part series, *Military Effectiveness* (1998), and the later *Military Innovation in the Interwar Period* are very fine examinations of military innovation and organizational change. See Murray and Millett.

40. Robert H. Scales Jr., *Future Warfare Anthology* (Carlisle, PA: U.S. Army War College, 2000), 43.

41. Hoffman, *Mars Adapting*, 273.

42. Bensahel and Barno, 270.

43. Rashed Zaman, "Kautilya: The Indian Strategic Thinker and Indian Strategic Culture," *Comparative Strategy* 25 (2006): 244.

44. Colin Gray, "Comparative Strategic Culture," *Parameters* 14, no. 4 (Winter 1984): 28.

45. Gray, *The Future of Strategy*, 23.

46. Alvin Bernstein, MacGregor Knox, and Williamson Murray, *The Making of Strategy: Rulers, States, and War* (Cambridge: Cambridge University Press, 1994), 615.

47. MacGregor Knox, "Conclusion: Continuity and Revolution in the Making of Strategy," in Bernstein, Knox, and Murray, 615–21.

48. According to at least one account, more than two hundred U.S. airmen were killed when their aircraft were shot down during missions near or over the Soviet Union during the Cold War. Paul Glenshaw, "Secret Casualties of the Cold War," *Air and Space Magazine* (December 2017), https://www.airspacemag.com/history-of-flight/secret-casualties-of-the-cold-war-180967122/.

49. This definition is from DOD, *Joint Operations* (Washington, DC: DOD, 17 January 2017), A3.

50. Lawrence Freedman, "Beyond Surprise Attack," *Parameters* 47, no. 2 (Summer 2017): 7.

51. Sun Tzu, 134–35.

52. Operation Bagration was a large Russian military operation in 1944. Through the use of deception and mass, the Russians inflicted a large defeat on the German army.

53. John Latimer, *Deception in War* (London: John Murray Publishers, 2001), 6.

54. J. F. C. Fuller, *The Foundations of the Science of War* (London: Hutchinson and Co. Publishers, 1926), 272–73.

55. Wohlstetter, viii.

56. Australian Army, *LWD1: The Fundamentals of Land Power* (Canberra: Australian Army, 2017), 33.

57. Bellingcat is an investigative journalism site based in England. Forensic Architecture is a research company based in London. For an example of their work on the Ukraine war, see their analysis of the battle of Ilovaisk at https://forensic-architecture.org/investigation/the-battle-of-ilovaisk.

58. Aric Toler, "Ukrainian Hackers Leak Russian Interior Ministry Docs with Evidence of Russian Invasion," *Global Voices*, 13 December 2014, https://globalvoices.org/2014/12/13/ukrainian-hackers-leak-russian-interior-ministry-docs-with-evidence-of-russian-invasion/.

59. Adamsky, *The Culture of Military Innovation*, 131–33.

60. Dima Adamsky, *Cross-Domain Coercion: The Current Russian Art of Strategy*, Proliferation Papers 54 (Paris: Institut Francais des Relations Internationales, 2015).

61. U.S. Senate Select Committee on Intelligence, *The Intelligence Community Assessment: Assessing Russian Activities and Intentions in Recent U.S. Elections, Summary of Initial*

*Findings*, July 3, 2018; Robert Mueller, *Report on the Investigation into Russian Interference in the 2016 Presidential Election*, vol. 1 (Washington, DC: Department of Justice, March 2019).

62. Michael Kofman, "The Moscow School of Hard Knocks: Key Pillars of Russian Strategy," *War on the Rocks*, 21 November 2019, https://warontherocks.com/2019/11/the-moscow-school-of-hard-knocks-key-pillars-of-russian-strategy-2/.

63. Dara Massicot, "Anticipating a New Russian Military Doctrine in 2020: What It Might Contain and Why It Matters," *War on the Rocks*, 9 September 2019, https://warontherocks.com/2019/09/anticipating-a-new-russian-military-doctrine-in-2020-what-it-might-contain-and-why-it-matters/.

64. Valery Gerasimov, "The Development of Military Strategy under Contemporary Conditions: Tasks for Military Science," *Military Review* (November 2019).

65. Adamsky, *Cross-Domain Coercion*, 9.

66. Massicot.

67. Kofman.

68. David Kilcullen, *The Dragons and the Snakes: How the Rest Learned to Fight the West* (Oxford: Oxford University Press, 2020), 209.

69. Clausewitz, 141.

70. Gray, *Modern Strategy*, 42–43.

71. Robert Leonhard, *Fighting by Minutes: Time and the Art of War* (KC Publishing, Kindle version, 2017), 13–16.

72. Leonhard, 13, 87.

73. Defense Advanced Research Projects Agency, "DARPA Tiles Together a Vision of Mosaic Warfare," https://www.darpa.mil/work-with-us/darpa-tiles-together-a-vision-of-mosiac-warfare.

74. Benjamin Jensen and John Paschkewitz, "Mosaic Warfare: Small and Scalable Are Beautiful," *War on the Rocks*, 23 December 2019, https://warontherocks.com/2019/12/mosaic-warfare-small-and-scalable-are-beautiful/.

75. Jeffrey Engstrom, *Systems Confrontation and System Destruction Warfare* (Santa Monica, CA: RAND Corporation, 2018), 15.

76. Michael Dahm, "Chinese Debates on the Military Utility of Artificial Intelligence," *War on the Rocks*, 5 June 2020, https://warontherocks.com/2020/06/chinese-debates-on-the-military-utility-of-artificial-intelligence/.

77. The concept of hyperwar is explored in Amir Husain, ed., *Hyperwar: Conflict and Competition in the AI Century* (Austin, TX: Spark Cognition Press, 2018).

78. This definition is the taken from U.S. Army, *Army Publication 3-90, Offense and Defense* (Washington, DC: Headquarters, Department of the Army, July 2019), 3-2.

79. Robert H. Scales Jr., *Yellow Smoke: The Future of Land Warfare for America's Military* (Oxford, UK: Rowman and Littlefield Publishers, 2003), 170.

80. Scales, *Future Warfare Anthology*, 4.

81. Williamson Murray, *Military Adaptation in War: With Fear of Change* (Cambridge: Cambridge University Press, 2011), 220–40.

82. "Democracies Need to Re-learn the Art of Deception," *The Economist*, 16 December 2020.

83. DOD, *Soviet Military Power 1990* (Washington, DC: DOD, 1990), 52; David Miller, *The Cold War: A Military History* (London: John Murray, 1999), 424.

84. Scott Boston et al., *Assessing the Conventional Force Imbalance in Europe: Implications for Countering Russian Local Superiority* (Santa Monica, CA: RAND Corporation, 2018), 3–5.

85. Boston et al., 4–6.

86. Anthony Cordesman, *U.S. Competition with China and Russia: The Crisis-Driven Need to Change U.S. Strategy* (Washington, DC: Center for Strategic and International Studies, May 2020).

87. International Institute for Strategic Studies, *The Military Balance 2020* (London: International Institute for Strategic Studies, 2020), https://www.iiss.org/publications/the-military-balance/military-balance-2020-book.

88. T. X. Hammes, "The Future of Warfare: Small, Many, Smart vs. Few and Exquisite?" *War on the Rocks*, 16 July 2014, https://warontherocks.com/2014/07/the-future-of-warfare-small-many-smart-vs-few-exquisite/.

89. C. Xu, "Intelligent War: Where Is the Change?" *Military Forum*, 21 January 2020, http://www.81.cn/jfjbmap/content/2020-01/21/content_252681.htm.

90. Ross Babbage, *Stealing a March: Chinese Hybrid Warfare in the Indo-Pacific*, vol. 1 (Washington, DC: Center for Strategic and Budgetary Assessments, 2019), 38–39.

91. Babbage, *Stealing a March*, 46.

92. Qiao and Wang, 19.

93. Qiao and Wang, 43.

94. Qiao and Wang, 41.

95. Babbage, *Stealing a March,* 69–75.

96. McDermott.

97. Qiao and Wang, 179.

98. Clausewitz, 78.

99. "The Metaverse Is Coming," *The Economist*, 1 October 2020, https://www.economist.com/technology-quarterly/2020/10/01/the-metaverse-is-coming.

100. Defense Science Board, *Counter Autonomy: Executive Summary* (Washington, DC: DOD, September 2020), 4.

101. Payne, "Artificial Intelligence: A Revolution in Strategic Affairs?" 23.

102. Different forms of human-machine command and control relationships are explained in Paul Scharre and Michael Horowitz, *An Introduction to Autonomy*

*in Weapon Systems* (Washington, DC: Center for a New American Security, February 2015), 6. Also Michael Ryan, "Extending the Intellectual Edge with Artificial Intelligence," *Australian Journal of Defence and Strategic Studies* 1, no. 1 (2019): 35–37.

103. The interaction of civilian leaders and senior military commanders is examined in Cohen, *Supreme Command: Soldiers, Statesmen, and Leadership in Wartime*.

104. Allan Millett and Williamson Murray, eds., *Military Effectiveness*, vol. 1, *The First World War* (Cambridge: Cambridge University Press, 2010), 6–12.

105. Millett and Murray, *Military Effectiveness*, 4–6.

106. Max Tegmark, *Life 3.0: Being Human in the Age of Artificial Intelligence* (New York: Alfred Knopf, 2017).

107. Forrest Morgan et al., *Military Applications of Artificial Intelligence: Ethical Concerns in an Uncertain World* (Santa Monica, CA: RAND Corporation, 2020), xiii.

108. Husain, 43.

109. Dr. Emily Goldman, speech at U.S. Naval War College seminar, Newport, RI, 7 August 2019.

110. This definition is based on that used by U.S. Department of Defense, *Joint Publication 3-13, Information Operations* (Washington, DC: Joint Chiefs of Staff, 2014), ix-x.

111. Sarah Kreps, "The Shifting Chessboard of International Influence Operations," Brookings Institution, 22 September 2020, https://www.brookings.edu/techstream /the-shifting-chessboard-of-international-influence-operations/.

112. Sorough Vosoughi, Deb Roy, and Sinan Aral, "The Spread of True and False News Online," *Science*, 9 March 2018, https://science.sciencemag.org/content/359 /6380/1146.

113. Babbage, *Winning without Fighting*, 29–34.

114. Kilcullen, 119.

115. Seely, 50–59. Chinese activities are described in multiple publications. Among these are the annual U.S. military report to Congress on China, *Military and Security Developments Involving the People's Republic of China 2019*; Japan's National Institute for Defense Studies report, *China Security Report 2019: China's Strategy for Reshaping the Asian Order and its Ramifications* (Tokyo: Ministry of Defense, 2019); and China's most recent national defense white paper, *China's National Defense in the New Era* (Beijing: Government of China, 24 July 2019). Useful analyses include: Elsa Kania and Peter Wood, "Major Themes in China's 2019 National Defense White Paper," *China Brief*, 31 July 2019, https://jamestown.org/program/major-themes-in-chinas -2019-national-defense-white-paper/; and Meia Nouwens, "China's 2019 Defence White Paper: Report Card on Military Reform," International Institute of Strategic Studies blog, 26 July 2019, https://www.iiss.org/blogs/analysis/2019/07/china-2019 -defence-white-paper.

116. W. Yang, "Exploring the Way to Win in Intelligent Warfare in Change and Invariance," *Military Forum*, 22 October 2019, http://www.81.cn/jfjbmap/content/2019-10/22/content_245810.htm.

117. Sean McFate, *The New Rules of War: Victory in the Age of Durable Disorder* (New York: William Morrow, 2019), 42.

118. Thomas Rid, *Active Measures: The Secret History of Disinformation and Political Warfare* (New York: Farrar, Straus, and Giroux, 2020), 11.

119. Security Council for the Russian Federation, *National Security Strategy 2020* (Moscow: Security Council for the Russian Federation, 2009).

120. UK Ministry of Defence, *Information Advantage*, Joint Concept Note 2/18 (Swindon: Defence Concepts and Doctrine Centre, 2018), ii.

121. Commonwealth of Australia, *Inquiry into the Implications of the COVID-19 Pandemic for Australia's Foreign Affairs, Defence, and Trade* (Canberra: Joint Standing Committee on Foreign Affairs, Defence, and Trade, 2020), xix.

122. Stephan Fruehling, *Sovereign Defence Industry Capabilities, Independent Operations, and the Future of Australian Defence Strategy*, Centre of Gravity Paper 36 (Canberra: Australian National University, October 2017).

123. Roland Rajah, "The International Economy," in the *World after COVID-19* series, Lowy Institute, 9 April 2020, https://interactives.lowyinstitute.org/features/covid19/issues/international-economy/.

124. Charles Edel, "Winning Over Down Under," *American Purpose*, 3 May 2021, https://www.americanpurpose.com/articles/winning-over-down-under/.

125. Peter Jennings, "Party's Over for the Bullies of Beijing," *The Australian*, 23 May 2020.

126. James Rogers et al., *Breaking the China Supply Chain: How the "Five Eyes" Can Decouple from Strategic Dependency* (London: Henry Jackson Society, May 2020), 8, https://henryjacksonsociety.org/publications/breaking-the-china-supply-chain-how-the-five-eyes-can-decouple-from-strategic-dependency/.

127. Rogers, 20.

128. Fruehling, 8.

129. Hew Strachan, "The Changing Character of War," Lecture at the Graduate Institute of International Relations, Geneva, 9 November 2006, 2.

130. Edmund Burke et al., *People's Liberation Army Operational Concepts* (Santa Monica, CA: RAND Corporation, 2020), 12.

## Chapter 3. Institutions, Ideas, and Future Military Effectiveness

1. Frank Stockton, *The Great War Syndicate* (New York: Dodd, Mead and Company, 1889), 12–13.

2. Stockton, 16–17.

3. Stockton, 120–35.

4. O'Hanlon *Technological Change and the Future of Warfare*, 7–8.

5. Andrew Krepinevich, *The Military-Technical Revolution: A Preliminary Assessment* (Washington, DC: Center for Strategic and Budgetary Assessments, 2002), 4–5.

6. In the 1970s, Soviet academics and military officers began to write about a new idea, that of a military technical revolution. Krepinevich, I, 5.

7. Andrew Liaropoulos has explored revolutions in military affairs and analyzed the complexity of the issue and the difficulty of examining its social, military, and technological dimensions. Liaropoulos, 363–84.

8. DOD, *Summary of the 2018 National Defense Strategy of the United States of America* (Washington, DC: DOD, 2018), 4–5; U.S. Government, *Interim National Security Strategic Guidance.*

9. U.S. Army Training and Doctrine Command, "The U.S. Army in Multidomain Operations 2028" (Fort Eustis, VA: Training and Doctrine Command, December 2018).

10. David Goldfein, speech to the 2016 Air, Space, and Cyber Conference, 20 September 2016, https://www.af.mil/Portals/1/documents/csaf/Goldfein_Air_Force_Update_Sept_2016.pdf; Amy McCullough, "Goldfein's Multi-Domain Vision," *Air Force Magazine*, 29 August 2018, https://www.airforcemag.com/article/goldfeins-multi-domain-vision/.

11. Defense Advanced Research Projects Agency, "Strategic Technology Office Outlines Vision for Mosaic Warfare," 4 August 2017, https://www.darpa.mil/news-events/2017-08-04.

12. Brian Clark, Daniel Patt, and Harrison Schramm, *Mosaic Warfare* (Washington, DC: Center for Strategic and Budgetary Assessments, 2020), iv.

13. Tai Ming Cheung, ed., *Forging China's Military Might: A New Framework for Assessing Innovation* (Baltimore: Johns Hopkins University Press, 2014), 1.

14. John Costello and Joe McReynolds, *China's Strategic Support Force: A Force for a New Era*, China Strategic Perspectives 13 (Washington, DC: National Defense University Press, September 2018).

15. DOD, *Military and Security Developments Involving the People's Republic of China 2019*, 15.

16. Government of China, *China's National Defense in the New Era* (Beijing: Chinese Communist Party, 24 July 2019), 12–13.

17. Engstrom, *Systems Confrontation and System Destruction Warfare.*

18. Joel Wuthnow, "China's Inopportune Pandemic Assertiveness," *The Diplomat*, 10 June 2020, https://thediplomat.com/2020/06/chinas-inopportune-pandemic-assertiveness/.

19. UK Ministry of Defence, *Defence in a Competitive Age* (London: UK Ministry of Defence, March 2021), 5.

20. Risa Brooks and Elizabeth Stanley, eds., *Creating Military Power: The Sources of Military Effectiveness* (Stanford, CA: Stanford University Press, 2007), 2–10.

21. Brooks and Stanley, 10–13.

22. Stephen Biddle and Stephen Long, "Democracy and Military Effectiveness: A Deeper Look," *Journal of Conflict Resolution* 48, no. 4 (August 2004): 525–46.

23. Millett and Murray, *Military Effectiveness*, 2.

24. Brooks and Stanley, 9.

25. Australian Defence Force, *Joint Planning* (Canberra: Defence Publishing Service, 2014), 1–3.

26. U.S. Department of Defense, *Joint Publication 1, Doctrine for the Armed Forces of the United States* (Washington, DC: Joint Chiefs of Staff, 2017), 1–7.

27. Australian Defence Force, *Foundations of Military Doctrine* (Canberra: Defence Publishing Service, 2012), 23.

28. United Kingdom Ministry of Defence, *UK Defence Doctrine* (Swindon: Defence Concepts and Doctrine Centre, November 2014), 7–11.

29. Heuser, 27.

30. Strachan, *Clausewitz's* On War, 106.

31. Gray, *The Future of Strategy*, 23.

32. Lawrence Freedman, *Strategy: A History* (Oxford: Oxford University Press, 2013), xi.

33. Heuser, 27–28.

34. Howard, *The Causes of War*, 194.

35. Allan Millett and Williamson Murray, "Lessons of War," *National Interest* 14 (Winter 1988): 83–95.

36. Brooks and Stanley, 10–11.

37. Aspects of strategic effectiveness are explored in Millett and Murray, *Military Effectiveness*, 5–9.

38. Government of the United Kingdom, *The Integrated Operating Concept 2025* (London: Ministry of Defence, 30 September 2020).

39. Peter Singer and Emerson Brooking, *Like War: The Weaponization of Social Media* (New York: Houghton Mifflin Harcourt Publishing, 2018), 261–62.

40. UK Ministry of Defence, *Information Advantage*, ii.

41. Millett and Murray, *Military Effectiveness*, 5.

42. Gavin Williamson, "Modernising Defence Programme—Update, Written Statement," UK Parliament website, 19 July 2018, https://www.parliament.uk/business/publications /written-questions-answers-statements/written-statement/Commons/2018-07-19 /HCWS883/.

43. Paul Bracken, "Net Assessment: A Practical Guide," *Parameters* 36, no. 1 (Spring 2006): 96.

44. George Pickett, James Roche, and Barry Watts, "Net Assessment: A Historical Review," in *On Not Confusing Ourselves*, ed. Andrew Marshall, J. J. Martin, and Henry Rowen (Boulder, CO: Westview Press, 1991), 166–67.

45. Andrew Marshall, "Long-Term Competition with the Soviets: A Framework for Strategic Analysis," (Santa Monica: RAND, 1972), 12.

46. UK Ministry of Defence, *Global Strategic Trends: The Future Starts Today*; NIC, *Global Trends: Paradox of Progress*.

47. Peter Layton et al., *How to Mobilise Australia*, Centre of Gravity Paper (Canberra: Australian National University, July 2020).

48. Geoffrey Blainey, *The Causes of War* (New York: Free Press, 1973), 246–49.

49. Murray, *War, Strategy, and Military Effectiveness*, 84–89.

50. Peter Mansoor and Williamson Murray, eds., *The Culture of Military Organizations* (Cambridge: Cambridge University Press, 2019), 1.

51. Mansoor and Murray, 1–3.

52. Hoffman, *Mars Adapting*, 16.

53. Lawrence Doane, "It's Just Tactics: Why the Operational Level Is an Unhelpful Fiction and Impedes the Operational Art," *Small Wars Journal*, 24 September 2015, https://small warsjournal.com/jrnl/art/it's-just-tactics-why-the-operational-level-of-war-is-an -unhelpful-fiction-and-impedes-the-.

54. North Atlantic Treaty Organization, *Allied Joint Doctrine*, AJP-01 (Brussels: NATO Standardization Office, February 2017), 1–10.

55. Craig Dalton, *Systemic Operational Design: Epistemological Bump or Way Ahead for Operational Design?* (Fort Leavenworth, KS: School of Advanced Military Studies, 2006), 5.

56. Millett and Murray, *Military Effectiveness*, 18–19.

57. Hans Binnendijk and Richard Kugler, "Adapting Forces to a New Era: Ten Transforming Concepts," *Defense Horizons* 5 (Washington, DC: National Defense University, November 2001), 3.

58. Engstrom, 10–15.

59. Michael O'Hanlon provides a description of this approach in Michael O'Hanlon, *The Science of War* (Princeton: Princeton University Press, 2009), 63–140.

60. Millett and Murray, *Military Effectiveness*, 16–17.

61. U.S. Army, Army Doctrine Publication 3-90, *Offense and Defense* (Washington, DC: Headquarters Department of the Army, July 2019), 1-1–1-4.

62. Murray, *Military Adaptation in War*, 13.

63. Dupuy, 317.

64. Robert H. Scales Jr., "Tactical Art in Future Wars," *War on the Rocks*, 14 March 2019, https://warontherocks.com/2019/03/tactical-art-in-future-wars/.

65. U.S. Marine Corps, *A Concept for Distributed Operations* (Washington, DC: Headquarters U.S. Marine Corps, 25 April 2005), 1.

66. David Berger, *38th Commandants Planning Guidance* (Washington, DC: Headquarters U.S. Marine Corps, July 2019). Also, U.S. Marine Corps, *Force Design 2030* (Washington, DC: Headquarters U.S. Marine Corps, March 2020).

67. Amos Fox and Andrew Rossow, "Making Sense of Russian Hybrid Warfare: A Brief Assessment of the Russo-Ukrainian War," Land Warfare Paper 112 (Arlington, VA: Association of the United States Army, 2017), 5–7.

68. Millett and Murray, *Military Effectiveness*, 20.

69. B. A. Friedman, *On Tactics: A Theory of Victory in Battle* (Annapolis, MD: Naval Institute Press, 2017), 19–21.

70. Friedman, 143.

71. Rita McGrath, "Transient Advantage," *Harvard Business Review*, June 2013, https://hbr.org/2013/06/transient-advantage.

72. Peter Schwartz, *The Art of the Long View* (New York: Bantam Doubleday Dell Publishing Group, 1991), 221.

73. Gray, *Strategy and Defence Planning*, 175.

74. Charles Darwin, *The Origin of the Species: By Means of Natural Selection or The Preservation of Favored Races in the Struggle for Life* (New York: Random House reprint, 1998), 636–37.

75. David Buss et al., "Adaptations, Exaptations, and Spandrels," *American Psychologist* 53, no. 4 (June 1998): 534.

76. Darwin, 636–37.

77. Robert Axelrod and Michael Cohen, *Harnessing Complexity: Organizational Implications of a Scientific Frontier* (New York: Free Press, 2000), 32.

78. Anne-Marie Grisogono, "Success and Failure in Adaptation," paper presented to the Sixth International Conference on Complex Systems, Boston, 25–30 June 2006, 2.

79. Grisogono, "Success and Failure in Adaptation," 7.

80. Grant Hammond, *The Mind of War: John Boyd and American Security* (Washington, DC: Smithsonian Institution Press, 2001), 35.

81. John Boyd, "Destruction and Creation," unpublished paper, 3 September 1976.

82. John Boyd, "Patterns of Conflict: Warp XII," unpublished briefing. Quantico, VA: Alfred Gray Research Center Archives, 20 March 1978.

83. Ian Brown, *A New Conception of War* (Quantico, VA: Marine Corps University Press, 2018), 123.

84. Brown, 150–72.

85. Michael Ryan, "Implementing an Adaptive Approach in Non-Kinetic Counterinsurgency Operations," *Australian Army Journal* IV, no. 3 (Summer 2007): 125–40.

86. Ryan, "Implementing an Adaptive Approach," 132.

87. Cohen and Gooch, 54–55.

88. Cohen and Gooch, 26.

89. Anne-Marie Grisogono, "The State of the Art and the State of the Practice: The Implications of Complex Adaptive Systems Theory for C2," paper for the 2006 Command and Control Research and Technology Symposium San Diego, June 2006.

90. Peter Senge, *The Fifth Discipline* (London: Random House, 2006), 17.

91. Australian Army, *Adaptive Campaigning* (Canberra: Australian Army, 2006), 7.

92. Michael Ryan, "Success and Failure of an Adaptive Army," *Australian Army Journal* VI, no. 3 (2009): 21–32.

## Chapter 4. People in Future Competition and War

1. Eugene Sledge, *With the Old Breed: At Peleliu and Okinawa* (New York: Ballantine Books, 2007), 150–51.

2. Martin van Crevald, *Command in War* (Cambridge, MA: Harvard University Press, 1985), 9.

3. Clausewitz, 100–102.

4. The institutional versus occupational approach to the profession of arms was examined in detail by Charles Moskos in his studies of the military profession in the United States. Charles Moskos, "From Institution to Occupation: Trends in Military Organization," *Armed Forces and Society* 4, no. 1 (November 1977): 41–50.

5. Yuval Harari, *Homo Deus: A Brief History of Tomorrow* (London: Vintage Books, 2015), 319.

6. James Stavridis, Ervin Rokke, and Terry Pierce, "Crafting and Managing Effects: The Evolution of the Profession of Arms," *Joint Force Quarterly* 81 (2nd Quarter 2016): 5.

7. Also explored in Ryan, "Extending the Intellectual Edge with Artificial Intelligence," 23–40.

8. Colin Gray, *War, Peace, and Victory: Strategy and Statecraft for the Next Century* (New York: Simon and Schuster, 1990), 14–15.

9. The 2018 U.S. National Defense Strategy notes that "the homeland is no longer a sanctuary." DOD, *Summary of the 2018 National Defense Strategy of the United States of America*, 3. In 2000 the Australian Defence White paper noted that "Australia is a secure country thanks to our geography." This phrase was absent in the 2016 edition of the same document. Commonwealth of Australia, *Defence 2000: Our Future Defence Force* (Canberra: Department of Defence, 2000), 23.

10. Sun Tzu, 192.

11. Clausewitz, 194.

12. Antoine Jomini, *The Art of War* (Philadelphia: J. B. Lippincott and Co., 1862), 70.

13. Mark Harrison, "Resource Mobilization for World War II: The USA, UK, USSR, and Germany, 1939–1945," *Economic History Review* 41, no. 2 (1988): 171–92; Condoleezza Rice, "The Making of Soviet Strategy," in *Makers of Modern Strategy from Machiavelli to the Nuclear Age*, ed. Peter Paret, Gordon A. Craig, and Felix Gilbert (Princeton: Princeton University Press, 1986), 655; U.S. Department of Defense, *Joint Publication 4-05, Joint Mobilization Planning* (Washington, DC: Joint Chiefs of Staff, 2018).

14. Leonhard, 13–16.

15. United States Institute of Peace, *Providing for the Common Defense: The Assessment and Recommendations of the National Defense Strategy Commission* (Washington, DC: U.S. Institute of Peace, 2018), viii; Ian Morris, *Why the West Rules—for Now: The Patterns of History, and What They Reveal about the Future* (New York: Farrar, Straus, and Giroux, 2010).

16. McGrath.

17. This guided my approach to commanding the Australian Defence College and sharing ideas with partners overseas. Michael Ryan, "An Australian Intellectual Edge for Conflict and Competition in the 21st Century," Centre of Gravity Paper 48 (Canberra: Australian National University, March 2019), and "The Intellectual Edge: A Competitive Advantage for Future War and Strategic Competition," *Joint Force Quarterly* 96 (1st Quarter 2020): 6–11.

18. Ryan, "Extending the Edge with Artificial Intelligence," 29–30.

19. Ryan.

20. Fox, 240–41.

21. Frank Hoffman, "Healthy Skepticism about the Future of Disruptive Technology and Modern War," Foreign Policy Research Institute blog, 4 January 2019, https://www.fpri.org/article/2019/01/healthy-skepticism-about-the-future-of-disruptive-technology-and-modern-war/.

22. These descriptions are taken from the career guide of the Australian Medical Association, available at https://ama.com.au/careers/doctors-training-and-career-advancement, and from Institution of Engineers Australia documents, available at https://www.engineersaustralia.org.au/sites/default/files/content-files/2016-12/career_development_guide_may_2014.pdf.

23. Australian Defence Force, *The Australian Joint Professional Military Education Continuum* (Canberra: Defence Publishing Service, 2019).

24. These descriptions build on my work and that of many others at the Australian Defence College in 2018 in the effort to define the required competencies and behaviors for military personnel at various stages of their careers, and to describe the joint

professional military education continuum and curricula to achieve these competencies and behaviors.

25. Charles Moskos and Frank Wood, eds., *The Military: More Than Just a Job?* (McLean, VA: Pergamon-Brassey's International Defense Publishers, 1988), 280.

26. For a detailed examination of the opportunities and challenges of future human-machine teaming in military organizations, see Ryan, *Human-Machine Teaming for Future Ground Forces*.

27. Van Crevald, 270.

28. Bernard Montgomery, *The Path to Leadership* (London: Collins Clear-type Press, 1961), 233.

29. The effective practice of mission command is essential to deal successfully with the chance, friction, and uncertainty of conflict. It also allows faster, more relevant decision-making in complex, volatile environments. Australian Army, *LWD1: The Fundamentals of Land Power*, 34.

30. Nick Bostrum, *Superintelligence: Paths, Dangers, Strategies* (Oxford: Oxford University Press, 2014).

31. DOD, *Joint Operations*, I-12.

32. DOD, *Joint Operations*, I-13.

33. Millett and Murray, *Military Effectiveness*, 15.

34. Measures of effectiveness drawn from the work of Millett and Murray, *Military Effectiveness*, 13–16.

35. Heuser, 488, 499.

36. William Rapp, "Civil-Military Relations: The Role of Military Leaders in Strategy Making," *Parameters* 45, no. 3 (Autumn 2015): 16.

37. Millett and Murray, *Military Effectiveness*, 6–12.

38. Clausewitz, 77.

39. Stephen Day, *Thoughts on Generalship: Lessons from Two Wars* (Canberra: Australian Army, 2015), 6.

40. Murray, *War, Strategy, and Military Effectiveness*, 6.

41. Murray, *War, Strategy, and Military Effectiveness*, 3.

42. This construct of institutional, educational, and technological initiatives has been a consistent theme in my articles and presentations on adapting our education and training for the challenges of the twenty-first century. See Ryan, "An Australian Intellectual Edge for Conflict and Competition in the 21st Century."

43. Adamsky, *The Culture of Military Innovation*, 142.

44. Richard Rumelt, *Good Strategy, Bad Strategy* (London: Profile Books, 2011).

45. John Setear et al., *The Army in a Changing World: The Role of Organizational Vision* (Santa Monica, CA: RAND Corporation, 1990), 68.

46. For the Singaporean approach, see https://www.mindef.gov.sg/oms/imindef/mindef
_websites/atozlistings/saftimi/units/ocs/ocs_journey.html.

47. Government of China, *Thirteenth Five-Year Science and Technology Military-Civil
Fusion Development Special Plan*, 26 September 2017, http://www.aisixiang.com/data
/106161.html.

48. Elsa Kania, "In Military-Civil Fusion, China Is Learning Lessons from the United
States and Starting to Innovate," *Strategy Bridge*, 27 August 2019, https://thestrategybridge
.org/the-bridge/2019/8/27/in-military-civil-fusion-china-is-learning-lessons-from
-the-united-states-and-starting-to-innovate.

49. UK Ministry of Defence, *Global Strategic Trends: The Future Starts Today*.

50. NIC, *Global Trends: Paradox of Progress*.

51. "Hollywood: The Pentagon's New Advisor," *BBC World News*, 2003, http://news.bbc.co.uk
/2/hi/programmes/panorama/1891196.stm.

52. Canadian Army, Directorate of Land Strategic Concepts, *Crisis in Zefra* (Kingston,
ON: Department of National Defence, 2005).

53. August Cole, ed., *War Stories from the Future* (Washington, DC: Atlantic Council,
2015).

54. A description of the Perry Group is at Michael Ryan, "Science Fiction, JPME, and the
Australian Defence College," *The Forge*, https://theforge.defence.gov.au/publications
/science-fiction-jpme-and-australian-defence-college.

55. Andrew Liptak, "The French Army Is Hiring Science Fiction Writers to Imagine Future
Threats," *The Verge*, 24 July 2019, https://www.theverge.com/2019/7/24/20708432
/france-military-science-fiction-writers-red-team.

56. Peter Singer and August Cole, *Ghost Fleet: A Novel of the Next World War* (Boston:
Houghton Mifflin and Harcourt, 2015); Peter Singer and August Cole, *Burn In*
(Boston: Houghton Mifflin and Harcourt, 2020).

57. August Cole, "Science Fiction and the Military Reader," *RUSI Journal* 162, no. 6
(December 2017): 60–64; Michael Ryan, "Why Reading Science Fiction Is Good
for Military Officers," *Grounded Curiosity*, 23 March 2016, https://grounded
curiosity.com/why-reading-science-fiction-is-good-for-military-officers/#.XyNhV
S2r1UM.

58. Chester Nimitz, speech to U.S. Naval War College, 1960, in Dale Rielage and
Trent Hone, "Counting the Cost of Learning: 'Learning War: The Evolution of
Fighting Doctrine in the U.S. Navy, 1895–1945,'" *Naval War College Review* 72, no.
2 (2019).

59. Edward Miller, *War Plan Orange: The U.S. Strategy to Defeat Japan 1897–1945*
(Annapolis, MD: Naval Institute Press, 1991), 168–69.

60. Allan Millett, "Assault from the Sea," in Murray and Millett, 72.

61. Millett, "Assault from the Sea," 74.

62. A detailed description of wargaming activities undertaken by the U.S. Naval War College is provided at https://usnwc.edu/Research-and-Wargaming/Wargaming.

63. The publicly available blog called "The War Room" is available at https://warroom .armywarcollege.edu.

64. MacGregor Knox, "Continuity and Revolution in Strategy," in Bernstein, Knox, and Murray, 615.

65. Michael Ryan, *The Ryan Review: A Study of the Army's Education, Training, and Doctrine Needs for the Future* (Canberra: Australian Army, April 2016), 46–47.

66. World Bank, *World Development Report 2019: The Changing Nature of Work* (Washington, DC: World Bank, 2019), 72.

67. Murray and Millett, 327.

68. The concept of insatiable curiosity about the profession of arms also featured in Australian Army, *Australian Army PME Strategy* (Canberra: Australian Army, October 2017), 6–7.

69. The capacity of institutions to "absorb" new technologies and capabilities is explored in Michael Horowitz, *The Diffusion of Military Power* (Princeton: Princeton University Press, 2010).

70. National Science and Technology Council, *Charting a Course for Success: America's Strategy for STEM Education* (Washington, DC: National Science and Technology Council, December 2018), v.

71. Paul Van Riper, *A Self-Directed Officer Study Program* (Carlisle, PA: U.S. Army War College, 1982), 1.

72. Charles White, *The Enlightened Soldier* (New York: Praeger, 1989).

73. Christine Redecker et al., *The Future of Learning: Preparing for Change* (Luxemburg: Joint Research Centre–Institute for Prospective Technological Studies, 2011), 29; World Bank.

74. Yuval Harari, "Why Technology Favors Tyranny," *The Atlantic* (October 2018): 64–70.

75. Peter Mansoor, "U.S. Army Culture 1973–2017," in Mansoor and Murray, 305.

76. Dan Marston has been a driving force in development of the Art of War program at the Australian Defence College and the Johns Hopkins University's School of Advanced International Studies.

77. These are some of the lessons shared among the military education community, as well as published on *The Forge*.

78. The theme of "access" was included in Australian Army, *Australian Army PME Strategy*, 6.

79. Cathy Downes, "Rapidly Evolving, Digitally-Enabled Learning Environments: Implications for Institutional Leaders, Educators and Students," in *Innovative Learning: A Key to National Security*, ed. Ralph Doughty, Linton Wells, and Theodore Hailes (Fort Leavenworth, KS: Army Press, 2015), 107–8, 124–25.

80. Established in 2018, *The Forge* is an Australian Defence Force site providing resources for self-study and unit-based professional military education. *The Cove* is an Australian Army–curated unclassified hub of professional material that engages learners outside of the work environment.

81. Rose Luckin and Wayne Holmes, *Intelligence Unleashed: An Argument for AI in Education* (London: Pearson and the UCL Knowledge Lab, 2016).

82. Explored in Michael Ryan, "Intellectual Preparation for Future War: How Artificial Intelligence Will Change Professional Military Education," *War on the Rocks*, 3 July 2018, https://warontherocks.com/2018/07/intellectual-preparation-for-future-war-how-artificial-intelligence-will-change-professional-military-education/.

83. Ryan, "Extending the Intellectual Edge with Artificial Intelligence," 32.

84. Ryan.

85. Justin Sanchez and Robbin Miranda, "Taking Neurotechnology into New Territory," in *Defense Advanced Research Projects Agency 1958–2018* (Tampa, FL: Faircount Media Group, 2018), 90–95.

86. Mary L. Cummings, "Technology Impedances to Augmented Cognition," *Ergonomics in Design* (Spring 2010): 25–27; Bijan Zakeri and Peter Carr, "The Limits of Synthetic Biology," *Trends in Biotechnology* (November 2014): 57–58.

87. Similar logic was used in developing the Australian Army professional military educations strategy in 2017. I am indebted to Lieutenant Colonel Tom McDermott in particular for his contributions.

88. Cohen, *The Big Stick*, 226.

89. Michael Ryan, "Mastering the Profession of Arms, Part III," *War on the Rocks*, 23 March 2017, https://warontherocks.com/2017/03/mastering-the-profession-of-arms-part-iii-competencies-today-and-into-the-future/.

## Conclusion

1. Michael Howard, *The Franco-Prussian War*, 2nd ed. (New York: Routledge, 2001), 212.

2. Howard, *The Franco-Prussian War*, 11.

3. Arden Bucholz, *Moltke and the German Wars, 1864–1871* (New York: Palgrave Publishers, 2001), 50–76; Reed Bonadonna, *Soldiers and Civilization: How the Profession of Arms Thought and Fought the Modern World into Existence* (Annapolis, MD: U.S. Naval Institute Press, 2017), 192–95.

4. Howard, *The Franco-Prussian War*, 1–2.

5. Qiao and Wang, 189.

6. Knox and Murray, 4; Marshall, "Some Thoughts on Military Revolutions," 4–5.

7. Harper.

8. Gray, *The Future of Strategy*, 6, 23.

9. Clausewitz, 87.

10. Freedman, *Strategy: A History*, xii.

11. Michael Howard, "Military Science in an Age of Peace," *RUSI Journal* 119, no. 1 (March 1974): 7.

12. Luciano Floridi, "Should We Be Afraid of AI?" *Aeon*, 9 May 2016, https://aeon.co/essays /true-ai-is-both-logically-possible-and-utterly-implausible.

13. Ryan, *Human-Machine Teaming for Future Ground Forces*, 8–11.

14. Qiao and Wang, 7.

15. Ryan, "An Australian Intellectual Edge for Conflict and Competition in the 21st Century," 4.

16. David Kilcullen, "Strategic Culture," in Mansoor and Murray, 51–52.

17. Clausewitz, 89.

18. Clausewitz, 87.

19. Howard, "Military Science in an Age of Peace," 7.

20. Clausewitz, 593.

21. Xi Jinping, "Report at the 19th National Congress of the Communist Party of China," *Xinhua*, 18 October 2017, http://www.xinhuanet.com/english/special/2017 -11/03/c_136725942.htm.

## Epilogue

1. Gray, *Modern Strategy*, 362.

2. Azar Gat, *Military Thought in the Nineteenth Century* (New York: Oxford University Press, 1992), 67.

3. Strachan, *Clausewitz's On War*, 142.

4. Clausewitz, 593.

5. Aaron Mehta, "AI Makes Mattis Question Fundamental Beliefs about War," *C4ISR Net*, 17 February 2018, https://www.c4isrnet.com/intel-geoint/2018/02/17/ai -makes-mattis-question-fundamental-beliefs-about-war/.

6. Hoffman, "Will War's Nature Change in the Seventh Military Revolution?" 31.

7. Daniel Dennett, "What Can We Do?" in *Possible Minds: 25 Ways of Looking at AI*, ed. John Brockman (New York: Penguin Press, 2019), 49.

8. Toby Walsh, *2062: The World That AI Made* (Carlton, VIC: La Trobe University Press, 2018), 31–33.

9. Strachan, "The Changing Character of War," 31.

# Bibliography

Adamsky, Dima. *Cross-Domain Coercion: The Current Russian Art of Strategy*. Proliferation Papers 54. Paris: Institut Francais des Relations Internationales, 2015.

———. *The Culture of Military Innovation: The Impact of Cultural Factors on the Revolution in Military Affairs in Russia, the U.S., and Israel*. Stanford, CA: Stanford Security Studies, 2010.

Angell, Norman. *The Great Illusion: A Study of the Relation of Military Power in Nations to their Economic and Social Advantage*. London: Putnam and Sons, 1911.

Arquilla, John. "The Big Kill." *Foreign Policy*, 3 December 2012. https://foreignpolicy.com/2012/12/03/the-big-kill/.

Australian Army. *Adaptive Campaigning*. Canberra: Australian Army, 2006.

———. *Australian Army PME Strategy*. Canberra: Australian Army, October 2017.

———. *LWD1: The Fundamentals of Land Power*. Canberra: Australian Army, 2017.

Australian Defence Force. *The Australian Joint Professional Military Education Continuum*. Canberra: Defence Publishing Service, 2019.

———. *Foundations of Military Doctrine*. Canberra: Defence Publishing Service, 2012.

———. *Joint Planning*. Canberra: Defence Publishing Service, 2014.

Axelrod, Robert, and Michael Cohen. *Harnessing Complexity: Organizational Implications of a Scientific Frontier*. New York: Free Press, 2000.

Babbage, Ross. *Stealing a March: Chinese Hybrid Warfare in the Indo-Pacific*, vol. 1. Washington, DC: Center for Strategic and Budgetary Assessments, 2019.

———. *Winning without Fighting: Chinese and Russian Political Warfare Campaigns and How the West Can Prevail*, vol. 1. Washington, DC: Center for Strategic and Budgetary Assessments, 2019.

Baldwin, Richard. *The Globotics Upheaval: Globalization, Robotics, and the Future of Work*. Oxford: Oxford University Press, 2019.

Bean, Charles. *Official History of Australia in the War of 1914–1918*, vol. 1: *The Story of ANZAC*. Sydney: Angus and Robertson Ltd., 1941.

Bensahel, Nora, and David Barno. *Adaptation under Fire: How Militaries Change in Wartime*. New York: Oxford University Press, 2020.

Berg, Jeremy, John Tymoczko, and Lubert Stryer. *Biochemistry*. New York: W. H. Freeman, 2002.

Berger, David. *38th Commandant's Planning Guidance*. Washington, DC: Headquarters U.S. Marine Corps, July 2019.

Berners-Lee, Tim. *Weaving the Web: The Original Design and Ultimate Destiny of the World Wide Web*. New York: Harper Business, 2000.

Bernstein, Alvin, MacGregor Knox, and Williamson Murray, eds. *The Making of Strategy: Rulers, States, and War*. Cambridge: Cambridge University Press, 1994.

Bernstein, William. *The Birth of Plenty: How the Prosperity of the Modern World Was Created*. New York: McGraw-Hill, 2004.

Biddle, Stephen. *Military Power: Explaining Victory and Defeat in Modern Battle*. Princeton: Princeton University Press, 2006.

Biddle, Stephen, and Stephen Long. "Democracy and Military Effectiveness: A Deeper Look." *Journal of Conflict Resolution* 48, no. 4 (August 2004): 525–46.

Biercuk, Michael, and Richard Fontaine. "The Leap into Quantum Technology: A Primer for National Security Professionals." *War on the Rocks*, 17 November 2017. https://waron therocks.com/2017/11/leap-quantum-technology-primer-national-security-professionals/.

Binnendijk, Hans, and Richard Kugler. "Adapting Forces to a New Era: Ten Transforming Concepts." *Defense Horizons* 5. Washington, DC: National Defense University Press, November 2001. https://ndupress.ndu.edu/Publications/Article/1215538/adapting -forces-to-a-new-era-ten-transforming-concepts/.

Blainey, Geoffrey. *The Causes of War*. New York: Free Press, 1973.

Blom, Philipp. *The Vertigo Years: Change and Culture in the West, 1900–1914*. London: Phoenix Books, 2009.

Bobbitt, Philip. *The Shield of Achilles: War, Peace, and the Course of History*. London: Penguin Books, 2002.

Bonadonna, Reed. *Soldiers and Civilization: How the Profession of Arms Thought and Fought the Modern World into Existence*. Annapolis, MD: Naval Institute Press, 2017.

Boot, Max. *War Made New: Technology, Warfare, and the Course of History, 1500 to Today*. New York: Gotham Books, 2006.

Boston, Scott, et al. *Assessing the Conventional Force Imbalance in Europe: Implications for Countering Russian Local Superiority*. Santa Monica, CA: RAND Corporation, 2018.

Bostrum, Nick. *Superintelligence: Paths, Dangers, Strategies*. Oxford: Oxford University Press, 2014.

Boyd, John. "Destruction and Creation." Unpublished paper, 3 September 1976.

———. "Patterns of Conflict: Warp XII." Unpublished briefing. Quantico, VA: Alfred Gray Research Center Archives, 20 March 1978.

Bracken, Paul. "Net Assessment: A Practical Guide." *Parameters* 36, no. 1 (Spring 2006): 90–100.

Brands, Hal, and Eric Edelman. *Why Is the World So Unsettled: The End of the Post–Cold War Era and the Crisis of the Global Order.* Washington, DC: Center for Strategic and Budgetary Assessments, 2017.

Braumoeller, Bear. *Only the Dead: The Persistence of War in the Modern Age.* New York: Oxford University Press, 2019.

Bricker, Darryl, and John Ibbitson. *Empty Planet: The Shock of Global Population Decline.* New York: Crown Publishers, 2019.

Brockman, John, ed. *Possible Minds: 25 Ways of Looking at AI.* New York: Penguin Press, 2019.

Brooks, Risa, and Elizabeth Stanley, eds. *Creating Military Power: The Sources of Military Effectiveness.* Stanford, CA: Stanford University Press, 2007.

Brown, Ian. *A New Conception of War.* Quantico, VA: Marine Corps University Press, 2018.

Bryn, Edward. *The Progress of Invention in the Nineteenth Century.* New York: Munn and Co. Publishers, 1900.

Bucholz, Arden. *Moltke and the German Wars, 1864–1871.* New York: Palgrave Publishers, 2001.

Burke, Edmund, et al. *People's Liberation Army Operational Concepts.* Santa Monica, CA: RAND Corporation, 2020.

Burrows, Matthew. *Global Risks 2035 Update: Decline of New Renaissance?* Washington, DC: Atlantic Council, 2019.

Buss, David, et al. "Adaptations, Exaptations, and Spandrels." *American Psychologist* 53, no. 4 (June 1998): 534.

Canadian Army, Directorate of Land Strategic Concepts. *Crisis in Zefra.* Kingston, ON: Department of National Defence, 2005.

Chandler, David. *The Campaigns of Napoleon: The Mind and Method of History's Greatest Soldier.* New York: Scribner, 1966.

Chase, Michael, et al. *China's Incomplete Military Transformation: Assessing the Weaknesses of the People's Liberation Army (PLA).* Santa Monica, CA: RAND Corporation, 2015.

Chatzky, Andrew, and James McBride. "China's Massive Belt and Road Initiative." Council on Foreign Relations Backgrounder. 28 January 2020. https://www.cfr.org /backgrounder/chinas-massive-belt-and-road-initiative.

Cheung, Tai Ming, ed. *Forging China's Military Might: A New Framework for Assessing Innovation.* Baltimore: Johns Hopkins University Press, 2014.

Cirillo, Pasquale, and Nassim Taleb. "On the Statistical Properties and Tail Risk of Violent Conflicts." *Physica A: Statistical Mechanics and its Applications* 452 (2016): 29–45.

Clark, Brian, Daniel Patt, and Harrison Schramm. *Mosaic Warfare.* Washington, DC: Center for Strategic and Budgetary Assessments, 2020.

Coats, Daniel R. "Worldwide Threat Assessment of the U.S. Intelligence Community." Statement for the Record to Senate Select Committee on Intelligence. Washington, DC: Office of the Director of National Intelligence, 29 January 2019.

Cohen, Eliot. *The Big Stick: The Limits of Soft Power and the Necessity of Military Force.* New York: Basic Books, 2016.

————. *Supreme Command: Soldiers, Statesmen, and Leadership in Wartime.* New York: Free Press, 2002.

Cohen, Eliot, and John Gooch. *Military Misfortunes: The Anatomy of Failure in War.* New York: Vintage Books, 1991.

Cole, August, ed. "Science Fiction and the Military Reader." *RUSI Journal* 162, no. 6 (December 2017): 60–64.

————. *War Stories from the Future.* Washington, DC: Atlantic Council, 2015.

Commonwealth of Australia. *2017 Foreign Policy White Paper.* Canberra: Department of Foreign Affairs and Trade, November 2017.

————. *Defence 2000: Our Future Defence Force.* Canberra: Department of Defence, 2000.

————. *Inquiry into the Implications of the COVID-19 Pandemic for Australia's Foreign Affairs, Defence, and Trade.* Canberra: Joint Standing Committee on Foreign Affairs, Defence, and Trade, 2020.

Commonwealth Scientific and Industrial Research Organization. *Australia's National Outlook 2019.* Canberra: Commonwealth Scientific and Industrial Research Organization, 2019.

Cordesman, Anthony. *U.S. Competition with China and Russia: The Crisis-Driven Need to Change U.S. Strategy.* Washington, DC: Center for Strategic and International Studies, May 2020.

Corum, James. *The Roots of Blitzkrieg.* Lawrence: University Press of Kansas, 1992.

Costello, John, and Joe McReynolds. *China's Strategic Support Force: A Force for a New Era.* China Strategic Perspectives 13. Washington, DC: National Defense University Press, September 2018.

Cronin, Audrey Kurth. *Power to the People: How Open Technological Innovation Is Arming Tomorrow's Terrorists.* New York: Oxford University Press, 2020.

Crump, Thomas. *The Age of Steam: The Power That Drove the Industrial Revolution.* London: Robinson, 2007.

————. *A Brief History of How the Industrial Revolution Changed the World.* London: Constable and Robinson Ltd., 2010.

Crutzen, Paul, and Eugene Stoermer. "The Anthropocene." *IGBP Newsletter* no. 41 (May 2000). http://www.igbp.net/download/18.316f18321323470177580001401/13763 83088452/NL41.pdf.

Cummings, Mary L. "Technology Impedances to Augmented Cognition." *Ergonomics in Design* (Spring 2010): 25–27.

Dahm, Michael. "Chinese Debates on the Military Utility of Artificial Intelligence." *War on the Rocks*, 5 June 2020. https://warontherocks.com/2020/06/chinese-debates-on-the-military-utility-of-artificial-intelligence/.

Dalton, Craig. *Systemic Operational Design: Epistemological Bump or Way Ahead for Operational Design?* Fort Leavenworth, KS: School of Advanced Military Studies, 2006.

Darwin, Charles. *The Origin of the Species: By Means of Natural Selection or the Preservation of Favored Races in the Struggle for Life.* New York: Random House reprint, 1998.

Day, Steven. *Thoughts on Generalship: Lessons from Two Wars*. Canberra: Australian Army, 2015.

Defense Advanced Research Projects Agency. "DARPA Tiles Together a Vision of Mosaic Warfare." https://www.darpa.mil/work-with-us/darpa-tiles-together-a-vision-of-mosiac -warfare.

———. "High Energy Liquid Laser Area Defense System." https://www.darpa.mil /program/high-energy-liquid-laser-area-defense-system.

———. "Strategic Technology Office Outlines Vision for Mosaic Warfare." 4 August 2017. https://www.darpa.mil/news-events/2017-08-04.

Defense Science Board. *Counter Autonomy: Executive Summary*. Washington, DC: DOD, September 2020.

Dempsey, Martin. *No Time for Spectators: The Lessons That Mattered Most from West Point to the West Wing*. Arlington, VA: Missionday Publishers, 2020.

Diamond, Larry. "Facing Up to the Democratic Recession." *Journal of Democracy* 26, no. 1 (January 2015): 141–55.

Doane, Lawrence. "It's Just Tactics: Why the Operational Level Is an Unhelpful Fiction and Impedes the Operational Art." *Small Wars Journal*, 24 September 2015. https://small warsjournal.com/jrnl/art/it's-just-tactics-why-the-operational-level-of-war-is-an-unhelpful -fiction-and-impedes-the-.

Dobbs, Michael. *One Minute to Midnight*. New York: Knopf, 2008.

Dobbs, Richard, James Manyika, and Jonathan Woetzal. *No Ordinary Disruption*. New York: Public Affairs, 2015.

Doudna, Jennifer, et al. "A Programmable Dual-RNA-Guided DNA Endonuclease in Adaptive Bacterial Immunity." *Science* (17 August 2012): 816–21.

Doughty, Ralph, Linton Wells, and Theodore Hailes, eds. *Innovative Learning: A Key to National Security*. Fort Leavenworth, KS: Army Press, 2015.

Dowling, Jonathan, and Gerard Milburn. "Quantum Technology: The Second Quantum Revolution." *The Royal Society* 361 (2003).

Dupont, Alan. *Mitigating the New Cold War: Managing U.S.-China Trade, Tech, and Geopolitical Conflict*. Sydney: Centre for Independent Studies, 2020.

Dupuy, Trevor. *The Evolution of Weapons and Warfare*. New York: Da Capo Press, 1984.

Edel, Charles. "Winning Over Down Under." *American Purpose*, 3 May 2021. https://www .americanpurpose.com/articles/winning-over-down-under/.

Engstrom, Jeffrey. *Systems Confrontation and System Destruction Warfare*. Santa Monica, CA: RAND Corporation, 2018.

Ferguson, Niall. *The War of the World: Twentieth-Century Conflict and the Descent of the West*. New York: Penguin Books, 2006.

Finney, Nathan, ed. *On Strategy: A Primer*. Fort Leavenworth, KS: Army University Press, 2020.

Finney, Nathan, and Tyrell Mayfield, eds. *Redefining the Modern Military: The Intersection of Profession and Ethics*. Annapolis, MD: Naval Institute Press, 2018.

FitzGerald, Alan, et al. *Economic Conditions Snapshot, March 2020: McKinsey Global Survey*. New York: McKinsey and Company, March 2020.

Floridi, Luciano. "Should We Be Afraid of AI?" *Aeon*, 9 May 2016. https://aeon.co/essays/true-ai-is-both-logically-possible-and-utterly-implausible.

Fox, Aimee. *Learning to Fight: Military Innovation and Change in the British Army, 1914–1918*. Cambridge: Cambridge University Press, 2017.

Fox, Amos, and Andrew Rossow. "Making Sense of Russian Hybrid Warfare: A Brief Assessment of the Russo-Ukrainian War." Land Warfare Paper 112. Arlington, VA: Association of the United States Army, 2017.

Freedberg, Sydney. "Meet the Army's Future Family of Robot Tanks: RCV." *Breaking Defense*, 9 November 2020. https://breakingdefense.com/2020/11/meet-the-armys-future-family-of-robot-tanks-rcv/.

Freedman, Lawrence. "Beyond Surprise Attack." *Parameters* 47, no. 2 (Summer 2017): 7.

———. *The Future of War: A History*. New York: Public Affairs Press, 2017.

———. *Strategy: A History*. Oxford: Oxford University Press, 2013.

Freedom House. *Freedom in the World 2019*. Washington, DC: Freedom House, 2019.

Freeman, Philip. *Alexander the Great*. New York: Simon and Schuster, 2011.

Friedman, B. A. *On Tactics: A Theory of Victory in Battle*. Annapolis, MD: Naval Institute Press, 2017.

Friedman, Thomas. *Thank You for Being Late*. New York: Macmillan, 2016.

Fruehling, Stephan. *Sovereign Defence Industry Capabilities, Independent Operations, and the Future of Australian Defence Strategy*. Centre of Gravity Paper. Canberra: Australian National University, October 2017.

Fry, Douglas, ed. *War, Peace, and Human Nature: The Convergence of Evolutionary and Cultural Views*. Oxford: Oxford University Press, 2013.

Fukuyama, Francis. "The End of History?" *The National Interest* no. 16 (Summer 1989).

———. *The End of History and the Last Man*. New York: Avon Books, 1992.

Fuller, J. F. C. *The Foundations of the Science of War*. London: Hutchinson and Co. Publishers, 1926.

———. *The Generalship of Alexander the Great*. Hertfordshire, UK: Wordsworth Editions Limited, 1998.

Future of Life Institute. "The Benefits and Risks of Artificial Intelligence." https://futureoflife.org/background/benefits-risks-of-artificial-intelligence/?cn-reloaded=1.

Garbee, Elizabeth. "This Is Not the Fourth Industrial Revolution." Slate.com, 29 January 2016. https://slate.com/technology/2016/01/the-world-economic-forum-is-wrong-this-isnt-the-fourth-industrial-revolution.html.

Gat, Azar. *Military Thought in the Nineteenth Century*. New York: Oxford University Press, 1992.

———. *War in Human Civilization*. Oxford: Oxford University Press, 2006.

Gerasimov, Valery. "The Development of Military Strategy under Contemporary Conditions: Tasks for Military Science." *Military Review* (November 2019). https://www.armyupress.army.mil/Portals/7/Army-Press-Online-Journal/documents/2019/Orenstein-Thomas.pdf.

Gettleman, Jeffrey, Hari Kumar, and Sameer Yasir. "Worst Clash in Decades on Disputed India-China Border Kills 20 Indian Troops." *New York Times*, 16 June 2020.

Gfoeller, Emmanuel. "Gap Warfare: The Case for a Shift in America's Strategic Mindset." *Real Clear Defense*, 28 May 2020. https://www.realcleardefense.com /articles/2020/05/28/gap_warfare_the_case_for_a_shift_in_americas_strategic _mindset_115325.html.

Gillies, James, and Robert Cailliau. *How the Web Was Born: The Story of the World Wide Web*. Oxford: Oxford University Press, 2000.

Glenshaw, Paul. "Secret Casualties of the Cold War." *Air and Space Magazine* (December 2017). https://www.airspacemag.com/history-of-flight/secret-casualties-of-the-cold -war-180967122/.

Goldfein, David. Speech to the 2016 Air, Space, and Cyber Conference, 20 September 2016. https://www.af.mil/Portals/1/documents/csaf/Goldfein_Air_Force_Update _Sept_2016.pdf.

Government of China. *China's National Defense in the New Era*. Beijing: Government of China, 24 July 2019.

———. *Thirteenth Five-Year Science and Technology Military-Civil Fusion Development Special Plan*. 26 September 2017. http://www.aisixiang.com/data/106161.html.

Government of Japan. *Defense of Japan 2017*. Tokyo: Government of Japan, 2017.

———. *Defense of Japan 2018*. Tokyo: Government of Japan, 2018.

Government of the Republic of France. *Defence and National Security Review 2017*. Paris: Government of the Republic of France, 2017.

Government of the United Kingdom. *The Integrated Operating Concept 2025*. London: Ministry of Defence, 30 September 2020.

Gray, Colin. "Comparative Strategic Culture." *Parameters* 14, no. 4 (Winter 1984): 26–33.

———. *The Future of Strategy*. Cambridge: Polity Press, 2015.

———. *Modern Strategy*. Oxford: Oxford University Press, 1999.

———. *Strategy and Defence Planning: Meeting the Challenge of Uncertainty*. Oxford: Oxford University Press, 2014.

———. "War—Continuity in Change, and Change in Continuity." *Parameters* 40, no. 2 (Summer 2010).

———. *War, Peace, and Victory: Strategy and Statecraft for the Next Century*. New York: Simon and Schuster, 1990.

Greenemeier, Larry. "20 Years after Deep Blue: How AI Has Advanced since Conquering Chess." *Scientific American*, 2 June 2017. https://www.scientificamerican.com/article /20-years-after-deep-blue-how-ai-has-advanced-since-conquering-chess/.

Grisogono, Anne-Marie. "The State of the Art and the State of the Practice: The Implications of Complex Adaptive Systems Theory for C2." Paper for the Command and Control Research and Technology Symposium, 2006.

———. "Success and Failure in Adaptation." Paper for the Sixth International Conference on Complex Systems, Boston, 25–30 June 2006.

Hafner, Katie, and Matthew Lyon. *Where Wizards Stay Up Late: The Origins of the Internet*. New York: Simon and Schuster, 1998.

Hammes, T. X. "Cheap Technology Will Challenge U.S. Tactical Dominance." *Joint Force Quarterly* 81 (2016): 76–85.

———. "The Future of Warfare: Small, Many, Smart vs. Few and Exquisite?" *War on the Rocks*, 16 July 2014. https://warontherocks.com/2014/07/the-future-of -warfare-small-many-smart-vs-few-exquisite/.

———. "3D Printing Will Disrupt the World in Ways We Can Barely Imagine." *War on the Rocks*, 28 December 2015. https://warontherocks.com/2015/12/3-d-printing-will -disrupt-the-world-in-ways-we-can-barely-imagine/.

Hammond, Grant. *The Mind of War: John Boyd and American Security*. Washington, DC: Smithsonian Institution Press, 2001.

Hanson, Victor Davis. *The Western Way of War: Infantry Battle in Classical Greece*. Berkeley: University of California Press, 2009.

Harari, Yuval. *Homo Deus: A Brief History of Tomorrow*. London: Vintage Books, 2015.

———. "Why Technology Favors Tyranny." *The Atlantic* (October 2018): 64–70.

Harper, Kyle. "The Coronavirus Is Accelerating History Past the Breaking Point." *Foreign Policy*, 6 April 2020. https://foreignpolicy.com/2020/04/06/coronavirus-is -accelerating-history-past-the-breaking-point/.

Harrison, Mark. "Resource Mobilization for World War II: The USA, UK, USSR, and Germany, 1939–1945." *Economic History Review* 41, no. 2 (1988).

Heginbotham, Eric, et al. *The U.S.-China Military Scorecard*. Santa Monica, CA: RAND Corporation, 2017.

Heuser, Beatrice. *The Evolution of Strategy: Thinking War from Antiquity to the Present*. Cambridge: Cambridge University Press, 2010.

Hockfield, Susan. *The Age of Living Machines: How Biology Will Build the Next Technology Revolution*. New York: W. W. Norton and Company, 2019.

Hoehn, Andrew, et al. *Discontinuities and Distractions: Rethinking Security for the Year 2040*. Santa Monica, CA: RAND Corporation, 2018.

Hoenig, Milton. "Artificial Intelligence: A Detailed Explainer, with a Human Point of View." *Bulletin of the Atomic Scientists*, 7 December 2018. https://thebulletin.org/2018/12/artificial -intelligence-a-detailed-explainer-with-a-human-point-of-view/.

Hoffman, Frank G. *Conflict in the 21st Century: The Rise of Hybrid Wars*. Washington, DC: Potomac Institute for Policy Studies, December 2007.

———. "Examining Complex Forms of Conflict: Gray Zone and Hybrid Challenges." *PRISM* 7, no. 4 (November 2018): 30–47.

———. "Healthy Skepticism about the Future of Disruptive Technology and Modern War." Foreign Policy Research Institute blog, 4 January 2019.

———. *Mars Adapting: Military Change during War*. Annapolis, MD: Naval Institute Press, 2021.

———. "Will War's Nature Change in the Seventh Military Revolution?" *Parameters* 47, no. 4 (Winter 2017–18): 23, 31.

Holmes, James, and Toshi Yoshihara. *Red Star Over the Pacific: China's Rise and the Challenge to U.S. Maritime Strategy*, 2nd ed. Annapolis, MD: Naval Institute Press, 2018.

Horne, Alistair. *To Lose a Battle*. London: Penguin Books, 1979.

Horowitz, Michael. *The Diffusion of Military Power*. Princeton: Princeton University Press, 2010.

Howard, Michael. *The Causes of War*, 2nd ed. Cambridge, MA: Harvard University Press, 1983.

———. *The Franco-Prussian War*, 2nd ed. New York: Routledge, 2001.

———. "Military Science in an Age of Peace." *RUSI Journal* 119, no. 1 (March 1974): 3–11.

Huntington, Samuel. *The Soldier and the State: The Theory and Politics of Civil-Military Relations*. Cambridge, MA: the Belknap Press, 1957.

Husain, Amir, ed. *Hyperwar: Conflict and Competition in the AI Century*. Austin, TX: Spark Cognition Press, 2018.

Institute for Security and Development Policy, *Made in China 2025: Backgrounder 2018*. Stockholm: Institute for Security and Development Policy, June 2018.

Intergovernmental Panel on Climate Change (IPCC). *Climate Change 2014: Synthesis Report Summary for Policy Makers*. Geneva: IPCC, 2015.

———. *Climate Change 2014: Synthesis Report. Contribution of Working Groups I, II, and III to the Fifth Assessment Report of the Intergovernmental Panel on Climate Change*. Geneva: IPCC, 2014.

International Institute for Strategic Studies. "Hypersonic Weapons and Strategic Stability." *Strategic Comments* 26 (March 2020).

———. *The Military Balance 2020*. London: International Institute for Strategic Studies, 2020. https://www.iiss.org/publications/the-military-balance/military -balance-2020-book.

Ioanes, Ellen. "China Steals U.S. Designs for New Weapons, and It's Getting Away with the Greatest Property Theft in Human History." *Business Insider*, 25 September 2019. https://www.businessinsider.com.au/esper-warning-china-intellectual-property-theft -greatest-in-history-2019-9?r=US&IR=T.

Ishimaru, Tota. *The Next World War*. London: Hurst and Blackett, 1936.

Jackson, Julian. *The Fall of France: The Nazi Invasion of 1940*. Oxford: Oxford University Press, 2004.

Janowitz, Morris. *The Professional Soldier: A Social and Political Portrait*. New York: Free Press, 1964.

Jee, Charlotte. "The First U.S. Trial of CRISPR Gene Editing in Cancer Patients Suggests the Technique Is Safe." *MIT Tech Review*, 7 February 2020. https://www.technologyreview .com/f/615157/the-first-us-trial-of-crispr-gene-editing-in-cancer-patients-suggests -the-technique-is-safe/.

Jennings, Peter. "Party's Over for the Bullies of Beijing." *The Australian*, 23 May 2020.

Jensen, Benjamin, and John Paschkewitz. "Mosaic Warfare: Small and Scalable Are Beautiful." *War on the Rocks*, 23 December 2019. https://warontherocks.com/2019 /12/mosaic-warfare-small-and-scalable-are-beautiful/.

Johnson, David. "An Army Caught in the Middle between Luddites, Luminaries, and the Occasional Looney." *War on the Rocks*, 19 December 2018. https://warontherocks .com/2018/12/an-army-caught-in-the-middle-between-luddites-luminaries-and-the -occasional-looney/.

————. *Fast Tanks and Heavy Bombers: Innovation in the U.S. Army, 1917–1945*. Ithaca: Cornell University Press, 2003.

Jomini, Antoine. *The Art of War*. Philadelphia: J. B. Lippincott and Co., 1862.

Jones, Bruce. "The New Geopolitics." Brookings Institution blog, 28 November 2017. https://www.brookings.edu/blog/order-from-chaos/2017/11/28/the-new-geopolitics/.

Kagan, Donald. "Intangible Interests and U.S. Foreign Policy." *Commentary*, April 1997. https://www.cs.utexas.edu/users/vl/notes/kagan.html.

————. *The Peloponnesian War*. London: Penguin Books, 2004.

Kagan, Robert. "The Twilight of the Liberal World Order." Brookings Institution blog, 24 January 2017. https://www.brookings.edu/research/the-twilight-of-the-liberal-world -order/.

Kahn, Laura. "A Crispr Future." *Bulletin of the Atomic Scientists*, 16 December 2015. https://thebulletin.org/2015/12/a-crispr-future/.

Kallenborn, Zachery. "Autonomous Drone Swarms as WMD." Modern War Institute, 28 May 2020. https://mwi.usma.edu/swarms-mass-destruction-case-declaring-armed-fully -autonomous-drone-swarms-wmd/.

Kania, Elsa. "China's Quantum Future: Xi's Quest to Build a High-Tech Superpower." *Foreign Affairs*, 26 September 2018. https://www.foreignaffairs.com/articles/china /2018-09-26/chinas-quantum-future.

————. "In Military-Civil Fusion, China Is Learning Lessons from the United States and Starting to Innovate." *Strategy Bridge*, 27 August 2019. https://thestrategybridge .org/the-bridge/2019/8/27/in-military-civil-fusion-china-is-learning-lessons-from-the -united-states-and-starting-to-innovate.

Kania, Elsa, and John Costello. *Quantum Hegemony: China's Ambitions and the Challenge to U.S. Innovation Leadership*. Washington, DC: Center for a New American Security, 2018.

Kania, Elsa, and Peter Wood. "Major Themes in China's 2019 National Defense White Paper." *China Brief*, 31 July 2019. https://jamestown.org/program/major-themes -in-chinas-2019-national-defense-white-paper.

Kasparov, Garry. *Deep Thinking: Where Machine Intelligence Ends and Human Creativity Begins*. London: John Murray Publishers, 2017.

"Kasparov vs. Deep Blue: The Match That Changed History." Chess.com, 12 October 2018. https://www.chess.com/article/view/deep-blue-kasparov-chess.

Keegan, John. *A History of Warfare*. New York: Alfred A. Knopf Inc., 1993.

Kennedy, Paul. *The Rise and Fall of the Great Powers: Economic Change and Military Conflict from 1500 to 2000*. New York: Random House, 1997, 191–92.

Kier, Elizabeth. *Imagining War: French and British Military Doctrine between the Wars*. Princeton: Princeton University Press, 1997.

Kilcullen, David. *The Dragons and the Snakes: How the Rest Learned to Fight the West*. Oxford: Oxford University Press, 2020.

Knox, MacGregor, and Williamson Murray, eds. *The Dynamics of Military Revolution, 1300–2050*. Cambridge: Cambridge University Press, 2001.

Kofman, Michael. "The Moscow School of Hard Knock: Key Pillars of Russian Strategy." *War on the Rocks*, 21 November 2019. https://warontherocks.com/2019/11/the-moscow -school-of-hard-knocks-key-pillars-of-russian-strategy-2/.

Kolbert, Elizabeth. "Peace in Our Time: Steven Pinker's History of Violence." *The New Yorker*, 26 September 2011. https://www.newyorker.com/magazine/2011/10/03/peace -in-our-time-elizabeth-kolbert.

KPMG. *Future State 2030: The Global Megatrends Shaping Governments*. Toronto: KPMG International, 2013.

Krepinevich, Andrew. *The Military-Technical Revolution: A Preliminary Assessment*. Washington, DC: Center for Strategic and Budgetary Assessments, 2002.

Kreps, Sarah. "The Shifting Chessboard of International Influence Operations." Brookings Institution, 22 September 2020. https://www.brookings.edu/techstream/the -shifting-chessboard-of-international-influence-operations/.

Kurzweil, Ray, and Chris Meyer. "Understanding the Accelerating Rate of Change." Kurzweilai.net blog, May 2003. https://www.kurzweilai.net/understanding-the -accelerating-rate-of-change.

Lacina, Bethany, and Nils Gleditsch. "Monitoring Trends in Global Combat: A New Dataset of Battle Deaths." *European Journal of Population* 21 (2005): 145–66.

———. "The Waning of War Is Real: A Response to Gohdes and Price." *Journal of Conflict Resolution* 57, no. 6 (2012): 1109–27.

Latimer, Jon. *Deception in War*. London: John Murray Publishers, 2001.

Layton, Peter, et al. "How to Mobilise Australia." Centre of Gravity Paper. Canberra: Australian National University, July 2020.

Leonhard, Robert. *Fighting by Minutes: Time and The Art of War*. KC Publishing (Kindle), 2017.

Levy, Steven. "What Deep Blue Tells Us about AI in 2017." *Wired Magazine*, 23 May 2017. https://www.wired.com/2017/05/what-deep-blue-tells-us-about-ai-in-2017/.

Liaropoulos, Andrew. "Revolutions in Warfare: Theoretical Paradigms and Historical Evidence: The Napoleonic and First World War Revolutions in Military Affairs." *The Journal of Military History* 70, no. 2 (April 2006): 363–84.

Lin, Jeffrey, and Peter Singer. "Drones, Lasers, and Tanks: China Shows Off Its Latest Weapons." *Popular Science*, 27 February 2017. https://www.popsci.com/china-new -weapons-lasers-drones-tanks/.

Liptak, Andrew. "The French Army Is Hiring Science Fiction Writers to Imagine Future Threats." *The Verge*, 24 July 2019. https://www.theverge.com/2019/7/24/20708432 /france-military-science-fiction-writers-red-team.

Luckin, Rose, and Wayne Holmes. *Intelligence Unleashed: An Argument for AI in Education*. London: Pearson and the UCL Knowledge Lab, 2016.

Lutz, Wolfgang. "World Population Trends and the Rise of Homo Sapiens Literata." Working Paper 19-012. Laxenburg, Austria: International Institute for Applied Systems Analysis, December 2019. http://pure.iiasa.ac.at/id/eprint/16214/1/WP-19-012.pdf.

Lynn, John. *Battle: A History of Combat and Culture*. New York: Westview Press, 2003.

MacCurdy, Edward, ed. *The Notebooks of Leonardo Da Vinci*. Old Saybrook, CT: Konecky and Konecky, 1955.

Mahnken, Thomas, ed. *Competitive Strategies for the 21st Century: Theory, History, and Practice*. Stanford, CA: Stanford University Press, 2012.

—. *Technology and the American Way of War since 1945*. New York: Columbia University Press, 2008.

Mahnken, Thomas, Ross Babbage, and Toshi Yoshihara. *Countering Comprehensive Coercion: Strategies against Authoritarian Political Warfare*. Washington, DC: Center for Defense and Strategic Studies, 30 May 2018.

Malthus, Thomas. *An Essay on the Principle of Population as It Affects the Future Improvement of Society*. London: J. Johnson Publishers, 1798.

Mansoor, Peter, and Williamson Murray, eds. *The Culture of Military Organizations*. Cambridge: Cambridge University Press, 2019.

Marshall, Andrew. "Competitive Strategies: History and Background." Unpublished paper. Washington, DC: Department of Defense, 3 March 1988.

—. "Long-Term Competition with the Soviets: A Framework for Strategic Analysis." Santa Monica, CA: RAND Corporation, 1972.

—. "Some Thoughts on Military Revolutions (Second Version)." Memorandum for the Record. Washington, DC: Department of Defense, 23 August 1993.

Massicot, Dara. "Anticipating a New Russian Military Doctrine in 2020: What It Might Contain and Why It Matters." *War on the Rocks*, 9 September 2019. https://warontherocks.com/2019/09/anticipating-a-new-russian-military-doctrine-in-2020-what-it-might-contain-and-why-it-matters/.

Mastro, Oriana Skylar. "The Stealth Superpower: How China Hid Its Global Ambitions." *Foreign Affairs* 98, no. 1 (January/February 2019).

Mazarr Michael, et al. *Understanding the Emerging Era of International Competition: Theoretical and Historical Perspectives*. Santa Monica, CA: RAND Corporation, 2018.

McCullough, Amy. "Goldfein's Multi-Domain Vision." *Air Force Magazine*, 29 August 2018. https://www.airforcemag.com/article/goldfeins-multi-domain-vision/.

McDermott, Roger. "Gerasimov Unveils Russia's Strategy of Limited Actions." *Eurasia Daily Monitor*, 6 March 2019. https://jamestown.org/program/gerasimov-unveils-russias-strategy-of-limited-actions/.

McFate, Sean. *The New Rules of War: Victory in the Age of Durable Disorder*. New York: William Morrow, 2019.

McGrath, Rita. "Transient Advantage." *Harvard Business Review*, June 2013. https://hbr.org/2013/06/transient-advantage.

McKenna, Chris. "Our World Is Changing—But Not as Rapidly as People Think." World Economic Forum, 2 August 2018. https://www.weforum.org/agenda/2018/08/change-is-not-accelerating-and-why-boring-companies-will-win/.

Mehta, Aaron. "AI Makes Mattis Question Fundamental Beliefs about War." *C4ISR Net*, 17 February 2018. https://www.c4isrnet.com/intel-geoint/2018/02/17/ai-makes-mattis-question-fundamental-beliefs-about-war/.

Mensch, Gerhard. *Stalemate in Technology: Innovations Overcome the Depression*. Cambridge: Ballinger, 1979.

Miller, David. *The Cold War: A Military History*. London: John Murray, 1999.

—. *Defense 2045: Assessing the Future Security Environment and Implications for Defense Policy Makers*. Washington, DC: Center for Strategic and International Studies, September 2015.

Miller, Edward. *War Plan Orange: The U.S. Strategy to Defeat Japan, 1897–1945.* Annapolis, MD: Naval Institute Press, 1991.

Millett, Allan, and Williamson Murray. "Lessons of War." *National Interest* 14 (Winter 1988): 83–95.

———, eds. *Military Effectiveness*, vol. 1: *The First World War.* Cambridge: Cambridge University Press, 2010.

Montgomery, Bernard. *The Path to Leadership.* London: Collins Clear-type Press, 1961.

Moorehead, Alan. *Gallipoli.* Melbourne: Macmillan Company of Australia, 1989.

Morgan, Forrest, et al. *Military Applications of Artificial Intelligence: Ethical Concerns in an Uncertain World.* Santa Monica, CA: RAND Corporation, 2020.

Morris, Charles. *America's First Industrial Revolution: The Dawn of Innovation.* Philadelphia: Public Affairs, 2012.

Morris, Ian. *War: What Is It Good For? The Role of Conflict in Civilization from Primates to Robots.* London: Profile Books, 2014.

———. *Why the West Rules—for Now: The Patterns of History, and What They Reveal about the Future.* New York: Farrar, Straus, and Giroux, 2010.

Morrison, Wayne. *The Made in China 2025 Initiative: Economic Implications for the United States.* Washington, DC: Congressional Research Service, 12 April 2019.

Moskos, Charles. "From Institution to Occupation: Trends in Military Organization." *Armed Forces and Society* 4, no. 1 (November 1977): 41–50.

Moskos, Charles, and Frank Wood, eds. *The Military: More Than Just a Job?* McLean, VA: Pergamon-Brassey's International Defense Publishers, 1988.

Mueller, John. "War Has Almost Ceased to Exist." *Political Science Quarterly* 124, no. 2 (Summer 2009): 297–321.

Mueller, Robert. *Report on the Investigation into Russian Interference in the 2016 Presidential Election*, vol. 1. Washington, DC: Department of Justice, March 2019.

Murray, Williamson. *Military Adaptation in War: With Fear of Change.* Cambridge: Cambridge University Press, 2011.

———. *War, Strategy, and Military Effectiveness.* Cambridge: Cambridge University Press, 2011.

Murray, Williamson, and Wayne Hsieh. *A Savage War: A Military History of the Civil War.* Princeton: Princeton University Press, 2016.

Murray, Williamson, and Allan Millett, eds. *Military Innovation in the Interwar Period.* Cambridge: Cambridge University Press, 1996.

Muzalevsky, Roman. *Strategic Landscape 2050: Preparing the U.S. Military for New Era Dynamics.* Carlisle, PA: Strategic Studies Institute and U.S. Army War College Press, 2017.

National Institute for Defense Studies. *China Security Report 2019: China's Strategy for Reshaping the Asian Order and Its Ramifications.* Tokyo: Ministry of Defense, 2019.

National Intelligence Council. *Global Trends: Paradox of Progress.* Washington, DC: National Intelligence Council, January 2017.

———. *Global Trends 2010.* Washington, DC: National Intelligence Council, 1997.

———. *Mapping the Global Future.* Washington, DC: National Intelligence Council, 2004.

National Science and Technology Council. *Charting a Course for Success: America's Strategy for STEM Education.* Washington, DC: National Science and Technology Council, December 2018.

North Atlantic Treaty Organization. *Allied Joint Doctrine,* AJP-01. Brussels: NATO Standardization Office, February 2017.

Nouwens, Meia. "China's 2019 Defence White Paper: Report Card on Military Reform." International Institute of Strategic Studies blog, 26 July 2019. https://www.iiss.org /blogs/analysis/2019/07/china-2019-defence-white-paper.

Office of the Director of National Intelligence. *Annual Threat Assessment of the U.S. Intelligence Community.* Washington, DC: Office of the Director of National Intelligence, 9 April 2021.

O'Hanlon, Michael. *Forecasting Change in Military Technology, 2020–2040.* Washington, DC: Brookings Institution, September 2018.

———. *The Science of War.* Princeton: Princeton University Press, 2009.

———. *Technological Change and the Future of Warfare.* Washington, DC: Brookings Institution, 2000.

Overholt, William. *China's Crisis of Success.* Cambridge: Cambridge University Press, 2018.

Paret, Peter. *The Cognitive Challenge of War: Prussia 1806.* Princeton: Princeton University Press, 2009.

Paret, Peter, Gordon A. Craig, and Felix Gilbert, eds. *Makers of Modern Strategy from Machiavelli to the Nuclear Age.* Princeton: Princeton University Press, 1986.

Payne, Kenneth. "Artificial Intelligence: A Revolution in Strategic Affairs?" *Survival* 60, no. 5 (October-November 2018): 7–32.

———. *Strategy, Evolution, and War: From Apes to Artificial Intelligence.* Washington, DC: Georgetown University Press, 2018.

Pickett, George, James Roche, and Barry Watts. "Net Assessment: A Historical Review." In Andrew Marshall, J. J. Martin, and Henry Rowen, eds., *On Not Confusing Ourselves.* Boulder, CO: Westview Press, 1991.

Pinker, Steven. *The Better Angels of Our Nature: Why Violence Has Declined.* New York: Penguin Books, 2011.

Porter, Michael. *Competitive Strategy: Techniques for Analyzing Industries and Competitors.* New York: Free Press, 1980.

Qiao, Liang, and Xiangsui Wang. *Unrestricted Warfare.* Beijing: PLA Literature and Arts Publishing House, February 1999.

Rajah, Roland. "The International Economy." In *The World after COVID-19* series, Lowy Institute, 9 April 2020. https://interactives.lowyinstitute.org/features/covid19 /issues/international-economy/.

Rand, Lindsay, and Berit Goodge. "Information Overload: The Promise and Risk of Quantum Computing." *Bulletin of the Atomic Scientists,* 14 November 2019. https://thebulletin .org/2019/11/information-overload-the-promise-and-risk-of-quantum-computing/.

Rapp, William. "Civil-Military Relations: The Role of Military Leaders in Strategy Making." *Parameters* 45, no. 3 (Autumn 2015): 13–26.

Redecker, Christine, et al. *The Future of Learning: Preparing for Change.* Luxembourg: Joint Research Centre–Institute for Prospective Technological Studies, 2011.

Rid, Thomas. *Active Measures: The Secret History of Disinformation and Political Warfare*. New York: Farrar, Straus, and Giroux, 2020.

Rielage, Dale, and Trent Hone. "Counting the Cost of Learning: 'Learning War: The Evolution of Fighting Doctrine in the U.S. Navy, 1895–1945.'" *Naval War College Review* 72, no. 2 (2019): 1–4.

Robitzski, Dan. "U.S. Army Doubles Down on Directed Energy Weapons." *Futurism .com*, 2 August 2019. https://futurism.com/the-byte/army-directed-energy-weapons.

Roff, Heather. *Uncomfortable Ground Truths: Predictive Analytics and National Security*. Washington, DC: Brookings Institution, 2020.

Rogers, James, et al. *Breaking the China Supply Chain: How the "Five Eyes" Can Decouple from Strategic Dependency*. London: Henry Jackson Society, May 2020. https://henryjacksonsociety.org/publications/breaking-the-china-supply-chain-how-the -five-eyes-can-decouple-from-strategic-dependency/

Rosen, Stephen. *Winning the Next War: Innovation and the Modern Military*. Ithaca: Cornell University Press, 1991.

Rosen, William. *The Most Powerful Idea in the World*. New York: Random House, 2010.

Roy, Shubhajit. "India China Border Tension: Chinese Ambassador Acknowledges PLA Deaths." *The Indian Express*, 27 June 2020. https://indianexpress.com/article/india /india-china-border-tension-chinese-ambassador-acknowledges-pla-deaths-6478314/.

Rumelt, Richard. *Good Strategy, Bad Strategy*. London: Profile Books, 2011.

Russell, Stuart. *Human Compatible: AI and the Problem of Control*. London: Allen Lane, 2019.

Russian Federation. *National Security Strategy*. Moscow: Ministry of Foreign Affairs, 31 December 2015.

Ryan, Michael. "An Australian Intellectual Edge for Conflict and Competition in the 21st Century." Centre of Gravity Paper 48. Canberra: Australian National University, March 2019.

———. "Extending the Intellectual Edge with Artificial Intelligence." *Australian Journal of Defence and Strategic Studies* 1, no. 1 (2019): 35–37.

———. *Human Machine Teaming for Future Ground Forces*. Washington, DC: Center for Strategic and Budgetary Assessments, 2018.

———. "Implementing an Adaptive Approach in Non-Kinetic Counterinsurgency Operations." *Australian Army Journal* IV, no. 3 (2007): 125–40.

———. "The Intellectual Edge: A Competitive Advantage for Future War and Strategic Competition." *Joint Force Quarterly* 96 (1st Quarter 2020): 6–11.

———. "Intellectual Preparation for Future War: How Artificial Intelligence Will Change Professional Military Education." *War on the Rocks*, 3 July 2018. https://warontherocks .com/2018/07/intellectual-preparation-for-future-war-how-artificial-intelligence -will-change-professional-military-education/.

———. "Mastering the Profession of Arms, Part III." *War on the Rocks*, 23 March 2017. https://warontherocks.com/2017/03/mastering-the-profession-of-arms-part-iii -competencies-today-and-into-the-future/.

———. *The Ryan Review: A Study of the Army's Education, Training, and Doctrine Needs for the Future*. Canberra: Australian Army, April 2016.

————. "Science Fiction, JPME, and the Australian Defence College." *The Forge*. https://the forge.defence.gov.au/publications/science-fiction-jpme-and-australian-defence-college.

————. "Success and Failure of an Adaptive Army." *Australian Army Journal* VI, no. 3 (2009): 21–32.

————. "Why Reading Science Fiction Is Good for Military Officers." *Grounded Curiosity*, 23 March 2016. https://groundedcuriosity.com/why-reading-science-fiction -is-good-for-military-officers/#.XyNhVS2r1UM.

Ryan, Michael, and Therese Keane. "Biotechnology and Human Augmentation: Issues for National Security Practitioners." *Strategy Bridge*, 5 February 2019. https://thestrategybridge .org/the-bridge/2019/2/5/biotechnology-and-human-augmentation-issues-for-national -security-practitioners.

Sanchez, Justin, and Robbin Miranda. "Taking Neurotechnology into New Territory." In *Defense Advanced Research Projects Agency 1958–2018*. Tampa, FL: Faircount Media Group, 2018.

Sargent, John, and R. X. Schwartz. *3D Printing: Overview, Impacts, and the Federal Role*. Washington, DC: Congressional Research Service, 2 August 2019.

Scales Jr., Robert H. *Future Warfare Anthology*. Carlisle, PA: U.S. Army War College, 2000.

————. "Tactical Art in Future Wars." *War on the Rocks*, 14 March 2019. https://warontherocks .com/2019/03/tactical-art-in-future-wars/.

————. *Yellow Smoke: The Future of Land Warfare for America's Military*. Oxford: Rowman and Littlefield Publishers, 2003.

Scharre, Paul. "Are AI-Powered Killer Robots Inevitable?" *Wired*, 19 May 2020. https:// www.wired.com/story/artificial-intelligence-military-robots/.

————. *Army of None: Autonomous Weapons and the Future of War*. New York: W. W. Norton and Company, 2019.

Scharre, Paul, and Michael Horowitz. *An Introduction to Autonomy in Weapon Systems*. Washington, DC: Center for a New American Security, February 2015.

Schelling, Thomas. *Arms and Influence*. New Haven, CT: Yale University Press, 1966.

Schwab, Klaus. *The Fourth Industrial Revolution*. New York: Crown Business, 2016.

Schwartz, Peter. *The Art of the Long View*. New York: Bantam Doubleday Dell Publishing Group, 1991.

Seely, Robert. "Defining Contemporary Russian Warfare." *RUSI Journal* 162, no. 1 (2017): 50–59.

Senge, Peter. *The Fifth Discipline*. London: Random House, 2006.

Setear, John, et al. *The Army in a Changing World: The Role of Organizational Vision*. Santa Monica, CA: RAND Corporation, 1990.

Shadbolt, Nigel, and Roger Hampson. *The Digital Ape: How to Live (in Peace) with Smart Machines*. Brunswick, VIC: Scribe Publications, 2018.

Shirk, Susan. *Fragile Superpower: How China's Internal Politics Could Derail Its Peaceful Rise*. Oxford: Oxford University Press, 2007.

Simpkin, Richard. *Race to the Swift*. Delhi: Lancer Publishers, 1997.

Singer, Peter, and Emerson Brooking. *Like War: The Weaponization of Social Media*. New York: Houghton Mifflin Harcourt Publishing, 2018.

Singer, Peter, and August Cole. *Ghost Fleet: A Novel of the Next World War*. Boston: Houghton Mifflin and Harcourt, 2015.

———. *Burn In*. Boston: Houghton Mifflin and Harcourt, 2020.

Singh, Sushant. "Explained: If Soldiers on the LAC Were Carrying Arms, Why Did They Not Open Fire?" *The Indian Express*, 20 June 2020. https://indianexpress.com /article/explained/explained-if-soldiers-on-lac-were-carrying-arms-why-did-they-not -open-fire-6467324/.

Sledge, Eugene. *With the Old Breed: At Peleliu and Okinawa*. New York: Ballantine Books, 2007.

Smil, Vaclav. *Creating the Twentieth Century: Technical Innovations of 1867–1914 and Their Lasting Impact*. Oxford: Oxford University Press, 2005.

Snegovaya, Maria. *Putin's Information Warfare in Ukraine: Soviet Origins of Hybrid Warfare*. Russia Report 1. Washington, DC: Institute for the Study of War, September 2015.

Speier, Richard, et al. *Hypersonic Missile Nonproliferation: Hindering the Spread of a New Class of Weapons*. Santa Monica, CA: RAND Corporation, 2017.

Stanley, Timothy, and Alexander Lee. "It's Still Not the End of History." *The Atlantic*, 1 September 2014. https://www.theatlantic.com/politics/archive/2014/09/its-still-not-the -end-of-history-francis-fukuyama/379394/.

Stavridis, James, Ervin Rokke, and Terry Pierce. "Crafting and Managing Effects: The Evolution of the Profession of Arms." *Joint Force Quarterly* 81 (2nd Quarter 2016): 4–9.

Stefanick, Tom. "The State of U.S.-China Quantum Data Security Competition." Washington, DC: Brookings Institution, 18 September 2020.

Steffen, Will, et al. *Global Change and the Earth System: A Planet under Pressure*. Berlin: IGBP Book Series, 2004.

———. "The Trajectory of the Anthropocene: The Great Acceleration." *The Anthropocene Review* 2, no. 1 (2015): 82–84.

Stewart, Phil. "U.S. Studying India Anti-Satellite Weapons Test, Warns of Space Debris." *Reuters*, 28 March 2019.

Stockton, Frank. *The Great War Syndicate*. New York: Dodd, Mead, and Company, 1889.

Stoker, Donald, and Craig Whiteside. "Blurred Lines: Gray-Zone Conflict and Hybrid War—Two Failures of American Strategic Thinking." *Naval War College Review* 73, no. 1 (2020): 13–48.

Strachan, Hew. "The Changing Character of War." Lecture at the Graduate Institute of International Relations, Geneva, 9 November 2006.

———. *Clausewitz's* On War: *A Biography*. New York: Grove Press, 2007.

Strassler, Robert, ed. *The Landmark Thucydides: A Comprehensive Guide to the Peloponnesian War*. Trans. Richard Crawley. New York: Free Press, 1998.

Suciu, Peter. "China's Army Now Has Killer Robots: Meet the Sharp Claw." *National Interest* blog, 17 April 2020. https://nationalinterest.org/blog/buzz/chinas-army -now-has-killer-robots-meet-sharp-claw-145302.

Sun, Yun. "China's Strategic Assessment of the Ladakh Clash." *War on the Rocks*, 19 June 2020. https://warontherocks.com/2020/06/chinas-strategic-assessment-of-the-ladakh -clash/.

Tegmark, Max. *Life 3.0: Being Human in the Age of Artificial Intelligence*. New York: Alfred Knopf, 2017.

Toler, Aric. "Ukrainian Hackers Leak Russian Interior Ministry Docs with Evidence of Russian Invasion." *Global Voices*, 13 December 2014. https://globalvoices.org /2014/12/13/ukrainian-hackers-leak-russian-interior-ministry-docs-with-evidence -of-russian-invasion/.

Trafton, Anne. "Making Smart Materials with CRISPR." *MIT Tech Review*, 24 October 2019. https://www.technologyreview.com/s/614509/making-smart-materials-with-crispr/.

Trevithick, Joseph. "China Conducts Test of Massive Suicide Drone Swarm Launched from a Box on a Truck." *The Warzone*, 14 October 2020. https://www.thedrive.com/the -war-zone/37062/china-conducts-test-of-massive-suicide-drone-swarm-launched-from -a-box-on-a-truck.

———. "Huge Navy Unmanned-Focused Experiment Underway Featuring Live Missile Shoot and Super Swarms," *The Warzone*, 20 April 2021. https://www.thedrive.com /the-war-zone/40262/huge-navy-unmanned-focused-experiment-underway-featuring -live-missile-shoot-and-super-swarms.

Tymoczko, John, et al. *Biochemistry*. New York: W. H. Freeman, 2002.

Tzu, Sun. *The Art of War*. Trans. Ralph Sawyer. Boulder, CO: Westview Press, 1994.

United Kingdom, Ministry of Defence. *Defence in a Competitive Age*. London: Ministry of Defence, March 2021.

———. *Global Strategic Trends: The Future Starts Today*, 6th ed. Swindon: Development, Concepts and Doctrine Centre, 2018.

———. *Information Advantage*. Joint Concept Note 2/18. Swindon: Development, Concepts and Doctrine Centre, 2018.

———. *UK Defence Doctrine*. Swindon: Development, Concepts and Doctrine Centre, November 2014.

United Nations. *International Migration Report 2017*. New York: United Nations Department of Economic and Social Affairs, Population Division, 2017.

———. *World Migration Report 2020*. New York: UN Department of Economic and Social Affairs, Population Division, 2019.

———. *World Population Aging 2019*. New York: United Nations Department of Economic and Social Affairs, Population Division, 2020.

———. *World Population Prospects 2019: Highlights*. New York: United Nations Department of Economic and Social Affairs, Population Division, 2019.

———. *World Urbanization Prospects: The 2018 Revision*. New York: United Nations Department of Economic and Social Affairs, 16 May 2018.

United States Institute of Peace. *Providing for the Common Defense: The Assessment and Recommendations of the National Defense Strategy Commission*. Washington, DC: U.S. Institute of Peace, 2018.

Unwin, Tim. "5 Problems with the 4th Industrial Revolution." *ICTWorks*, 23 March 2019. https://www.ictworks.org/problems-fourth-industrial-revolution/#.X97v MC0Rokh.

U.S. Army. Army Doctrine Publication 3-90, *Offense and Defense*. Washington, DC: Headquarters Department of the Army, July 2019.

————. *The Operational Environment and the Changing Character of Warfare.* Fort Eustis, VA: Department of the Army, October 2019.

————. Training and Doctrine Command. "The U.S. Army in Multidomain Operations 2028." Fort Eustis, VA: Department of the Army, December 2018.

U.S. Department of Defense. *Doctrine for the Armed Forces of the United States.* Washington, DC: Joint Chiefs of Staff, 2017.

————. *Indo-Pacific Strategy Report.* Washington, DC: Department of Defense, June 2019.

————. *The Joint Force in a Contested and Disordered World.* Washington, DC: Joint Chiefs of Staff, 14 July 2016.

————. *Joint Publication 4-05, Joint Mobilization Planning.* Washington, DC: Joint Chiefs of Staff, 23 October 2018.

————. *Joint Operations.* Washington, DC: Department of Defense, 17 January 2017.

————. *Joint Publication 3-13, Information Operations.* Washington, DC: Joint Chiefs of Staff, 2014.

————. *Military and Security Developments Involving the People's Republic of China 2019.* Washington, DC: Office of the Secretary of Defense, 2019.

————. *National Defense Strategy of the United States of America.* Washington, DC: Department of Defense, 2017.

————. *Soviet Military Power 1990.* Washington, DC: Department of Defense, 1990.

————. *Summary of the 2018 National Defense Strategy of the United States of America.* Washington, DC: Department of Defense, 2018.

————. *Sustaining U.S. Global Leadership: Priorities for 21st-Century Defense.* Washington, DC: Department of Defense, 2012.

U.S. Government. *Interim National Security Strategic Guidance.* Washington, DC: White House, 2021.

U.S. Marine Corps. *A Concept for Distributed Operations.* Washington, DC: Headquarters U.S. Marine Corps, 25 April 2005.

————. *Competing.* Washington, DC: Headquarters U.S. Marine Corps, 14 December 2020.

————. *Force Design 2030.* Washington, DC: Headquarters U.S. Marine Corps, March 2020.

U.S. Senate Select Committee on Intelligence. *The Intelligence Community Assessment: Assessing Russian Activities and Intentions in Recent U.S. Elections, Summary of Initial Findings,* July 3, 2018.

Van Crevald, Martin. *Command in War.* Cambridge, MA: Harvard University Press, 1985.

Van Riper, Paul. *A Self-Directed Officer Study Program.* Carlisle, PA: U.S. Army War College, 1982.

Vergun, David. "Military Leaders Discuss Hypersonics, Supply Chain Vulnerabilities." *DoD News,* 21 February 2020.

Vermeer, Michael, and Evan Peet. *Securing Communications in the Quantum Computing Age.* Santa Monica, CA: RAND Corporation, 2020.

von Clausewitz, Carl. *On War.* Trans. Michael Howard and Peter Paret. Princeton: Princeton University Press, 1976.

Vosoughi, Soroush, Deb Roy, and Sinan Aral. "The Spread of True and False News Online." *Science* 359 (2018): 1146–51.

Walsh, Toby. *2062: The World That AI Made*. Carlton, VIC: La Trobe University Press, 2018.

Warner, Ben. "Pentagon Shifts Focus on Directed Energy Weapons Technology." U.S. Naval Institute blog, 5 September 2019. https://news.usni.org/2019/09/05/pentagon -shifts-focus-on-directed-energy-weapons-technology.

Weick, Karl, and Kathleen Sutcliffe. *Managing the Unexpected: Assuring High Performance in an Age of Complexity*. San Francisco: Jossey-Bass, 2001.

White, Charles. *The Enlightened Soldier*. New York: Praeger, 1989.

Williamson, Gavin. "Modernising Defence Programme—Update, Written Statement." UK Parliament website, 19 July 2018. https://www.parliament.uk/business/publications /written-questions-answers-statements/written-statement/Commons/2018-07-19 /HCWS883/.

Woetzel, Jonathan, et al. *People on the Move: Global Migration's Impact and Opportunity*. New York: McKinsey Global Institute, December 2016.

Wohlstetter, Roberta. *Pearl Harbor: Warning and Decision*. Stanford, CA: Stanford University Press, 1962.

World Bank. *World Development Report 2019: The Changing Nature of Work*. Washington, DC: World Bank, 2019.

Wuthnow, Joel. "China's Inopportune Pandemic Assertiveness." *The Diplomat*, 10 June 2020. https://thediplomat.com/2020/06/chinas-inopportune-pandemic-assertiveness/.

Xi Jinping. "Report at the 19th National Congress of the Communist Party of China." *Xinhua*, 18 October 2017. http://www.xinhuanet.com/english/special/2017–11/03 /c_136725942.htm.

Xu, C. "Intelligent War: Where Is the Change?" *Military Forum*, 21 January 2020. http://www.81.cn/jfjbmap/content/2020-01/21/content_252681.htm._

Yang, W. "Exploring the Way to Win in Intelligent Warfare in Change and Invariance." *Military Forum*, 22 October 2019. http://www.81.cn/jfjbmap/content/2019-10/22 /content_245810.htm.

Zak, Anatoly. "The Hidden History of the Soviet Satellite Killer." *Popular Mechanics*, 1 November 2013. https://www.popularmechanics.com/space/satellites/a9620/the -hidden-history-of-the-soviet-satellite-killer-16108970/.

Zakeri, Bijan, and Peter Carr. "The Limits of Synthetic Biology." *Trends in Biotechnology* (November 2014).

Zaman, Rashed. "Kautilya: The Indian Strategic Thinker and Indian Strategic Culture." *Comparative Strategy* 25 (2006): 244

Zubrin, Robert. *Entering Space: Creating a Spacefaring Civilization*. New York: Penguin Putnam Inc., 1999.

# Index

academia, strategic engagement and, 137, 191

*Active Measures* (Rid), 111

Adamsky, Dima, 79

adaptation and adaptive capacity: as AGI obstacle, 230; building, 12; continuous learning and, 198; elements, 161; future effectiveness and, 154–59, 225; future warfare and, 125–26; as human imperative, 168; of military institutions, 216–17; national security leaders and, 185; pace of change and, 129–30; systemic integration of technologies and, 153–54; training and education for, 196; working understanding of, 182–83. *See also* counteradaptation

*Adaptation under Fire* (Bensahel and Barno), 73

additive manufacturing, 53–54

Advanced Research Projects Agency Network (ARPANET), 20

aerospace, *Made in China 2025* and, 35

Age of Accelerations, 22

aging populations, 40, 59

AI extenders, 206. *See also* artificial intelligence

Air Force, U.S., 122, 202

aircraft, signature management by, 92

Alexander the Great, 8

algorithms. *See* artificial intelligence; human-machine integration; robotics

Allen, John, 107–8

alliances, 136, 138. *See also* joint operations

Anglo-American syndicate of war, 119

Anthropocene, definition, 55

"AR5 Synthesis Report: Climate Change 2014," 56

Archimedes, 49

Army, U.S.: adaptive capacity, 156; conservative approach to change, 72–73, 247n36; laser weapons and, 49; on new technologies, 121; "Robotic Warfare Battlefield Geometry," 45; School of Advanced Military Studies (SAMS), 201; Training and Doctrine Command, 24; "The U.S. Army in Multidomain Operations 2028," 122

Army War College, U.S., 195

*Art of the Long View, The* (Schwartz), 155

*Art of War, The* (Chinese treatise), 123

*Art of War, The* (Jomini), 170

*Art of War, The* (Machiavelli), 233n4

artificial general intelligence (AGI), 42, 229–30

285

advantage and, 171, 172; as contemporary driver of disruption, 28; disruptive, literacy in, 219–20; energy weapons, 49–50; engagement in the global PME ecosystem, 204; ethical and creative use of, 217–19; evolution of weapons systems and, 88; future warfare and, 125–26; hypersonics, 50–51; impacts, 60, 245n148; imperial expansion of European nations and, 241n83; improved access to joint education and training and, 202–3; informatization and intelligentization of warfare and, 86–87, 124; innovation in engagement and delivery, 204–5; leadership and literacy in, 177; literacy, intellectual edge and, 198–99; literacy, leadership development and, 182, 183; military revolution in (early 1990s) and, 120–22, 253n6; military robotics, 44–46; nature of war and, 227–31; new-age massed forces and, 97–98; ongoing developments in, military institutions and, 5–6, 9, 18–19, 20, 27; operational concepts and, 145; overview of new and evolved, 41–42; Prussian military and, 209–10; quantum technology, 46–47; Russia's warfare with Ukraine and, 78–80; space technology, 51–53; systemic integration of, 153–54; training and education for, 152–53; U.S. Army studies after Arab-Israeli War of, 121; of war, Industrial Revolutions and, 210–11. *See also* revolutions

Tegmark, Max, 229–30

tempo, controlling, for future conflicts, 89

terrorists, unpredictability of, 90

*Thank You for Being Late* (Friedman), 22–23, 55

think tanks, 137, 194–96

three-dimensional (3D) printing, 53–54, 83

Thucydides, 8

time, 82–83, 84–90, 171

training: adaptive training and education, 196; creativity, futures, and science fiction, 191–94; educational institutions as think tanks, 194–96; informal and institutional, 168; investment in education and, 173; joint by design not by qualification, 189–90; strategic design of, 187–89; strategic engagement, 190–91; systems, tactical level of war and, 152–53

twenty-first century trends: human-machine integration, 102–8; influence, evolving fight for, 108–12; integrated thinking and action, 98–102; mass, new forms of, 94–98; overview, 82–84; signatures, battle of, 90–94; sovereign resilience, 112–16; time, appreciation of, 84–90. *See also* future effectiveness

Ukraine, 34, 75, 77–79, 149

United Kingdom: Defence Concept Development Centre, 60–61; Development, Concepts and Doctrine Centre, 29, 30; joint operations and, 98; strategic dependency on China of, 114; on strategic level of war, 129. *See also* Britain

United Nations, 37, 39, 56

United States: cold but connected war with China and, 38; competition with Russia, 36–37; on military robotics regulations, 46; military technological revolution and, 120–22; on mosaic warfare, 86; Russian and Chinese strategies against, 33–34; space technology and, 52; strategic culture of, 65–66; strategic dependency on China of, 114; surprise as key principle of war for, 3

universities, civilian, strategic engagement and, 137, 191

unmanned systems, 96–97. *See also* autonomous systems; robotics

# About the Author

**Mick Ryan** is a major general in the Australian Army. A distinguished graduate of Johns Hopkins University School of Advanced International Studies, as well as the USMC Command and Staff College and USMC School of Advanced Warfare, he is a passionate advocate of professional education and lifelong learning. He has commanded at platoon, squadron, regiment, task force, and brigade level. Ryan has also led strategic planning organizations in the Australian Army and led several reform programs in the past decade. In January 2018, he assumed command of the Australian Defence College in Canberra, Australia.